Ali Ukasha

Practical Contour Compression Methods Using MATLAB

Ali Ukasha

Practical Contour Compression Methods Using MATLAB

Noor Publishing

Imprint

Cover image: www.ingimage.com

Publisher:
Noor Publishing
is a trademark of
Dodo Books Indian Ocean Ltd., member of the OmniScriptum S.R.L Publishing group
str. A.Russo 15, of. 61, Chisinau-2068, Republic of Moldova Europe
Printed at: see last page
ISBN: 978-620-2-79033-8

Practical Contour Compression Methods Using MATLAB

by

Ali Abdrhman M. Ukasha

Sebha University

Faculty of Engineering & Technology Sciences

Electrical & Electronic Engineering Department

2020

TABLE OF CONTENTS

	Page
Preface	4
Introduction	6
1. Matlab Basics Operations	**7**
1.1 Special Characters..	7
1.2 Arithmetic Operators	8
1.3 Relational operations	11
1.4 Logical operators	12
1.5 Commands and functions	14
1.6 Important Statements	20
1.7 Image Formats	23
1.8 Image Types	24
2. Contour Extraction Explanation	**33**
2.1 Introduction	33
2.2 Color & Binary Images	33
2.3 Contour extraction methods	39
2.3.1 Object contour following 'OCF' method	42
2.3.2 Multiple step contour extraction 'MSCE' method	43
2.3.3 Object-oriented contour extraction 'OCE' method	44
2.3.4 Single Step Parallel Contour Extraction 'SSPCE' Method	45
2.4 Contour Extraction Examples using Matlab	54
2.5 Edge detectors	60
3. Contour Compression Methods in Spatial Domain	64
3.1 Ramer Method	64

3.2 Tangent Method 78
3.3 Triangle Method 93
3.4 Trapezoid Method 101
3.5 Adaptive-Triangle Method 103
3.6 Experimental Results 107

4. Contour Compression Methods in Spectral Domain 115
4.1 Overvieew 115
4.2 Discrete Cosine Transform (DCT) 116
4.3 Contour Compression in Transform Domain 118
4.4 Contour compression using Zonal Sampling 123
4.5 Haar Wavelet Transform 127
4.6 Comparison between the fast transforms and time domain methods 130
4.7 Discrete Wavelet Transform (DWT) 132
4.8 Applied Measures 133
4.9 Results of the Experiments 134

REFRENCES **148**
APPENDIXES **152**

Preface

We are aim to serves two purposes: first to introduce readers the concepts of contour compression methods, and secondly to provide them with experience in modeling the theory and applications using the commonly used software tool MATLAB. This book is deals with contour extraction & compression methods in spatial & spectral domains.

MATLAB examples are given throughout the text. MATLAB is emphasized since it is now a widely accepted software tool very routinely used in signal processing. A sizeable portion of this book is based on the authors' own in-class presentations, as well as research in the area. Instructive problems and MATLAB assignments are included at the end of each Chapter to enhance even further the value of this book to its readers.

Written for undergraduate students, Lab Primer through MATLAB presents a number of techniques which help students carry out various experiments in the fields of contour image compression. A step-wise discussion of the programming procedure using MATLAB has been carried out in this book.

In this edition, the completely new material is on new methods for low- and high-level contour extraction, description and compression. We have also extended the book to use color and more modern techniques for object extraction and description especially those capitalize wavelets and DCT transform. We will address how to efficiently represent and compress image/contour signals. Topics to be covered include: DCT & Wavelet transforms, Low and high-pass filtering, morphological operations, noise removal, edge detection, image compression and object recognition.

About the Author

Ali Abdrhman M. Ukasha
A. A. M. Ukasha received his B.S.C. degree from the University of Garyounis, Libya, in 1986 and the M.Sc. degree from Poznan University of Technology, Poland, in 1998, and Ph.D. degree from AGH University of Science and Technology, Cracov, Poland, in 2007.

He was joined the Electronics and Computer Engineering Department at the University of Sebha in (1992-1995) as a Demonstrator, then he gone to continue his higher Education in Poland. He back to his department in 1999 as a Lecturer. He became Assistant Professor in 2010, Associate professor in 2014, and Professor in 2018.

He was working as an organizer of the higher studies program in his Department. He served as Chairman of the department from 2015 up to 2016. Recently, I am working as head of Quality performance office in the Faculty of Engineering and Technology Sciences, Sebha University, Libya. Dr. A. Ukasha published more than 40 papers at International Conferences and Journals in the fields of Telecommunications Engineering and Image Processing. He is a member of the technical program committee in the International conference of the Signal, Image and Speech Processing (SISP) and Editorial Review Board on Electronics and Communication Engineering, World Academy of Science, Engineering and Technology (WASET) organization. He is one of the Journal of Software Engineering & Intelligent Systems (JSEIS) Editorial Board. He joint as Conference technical committee at International Congress on Recent Development in Engineering and Technology (RDET-16). In addition, Prof. Ukasha is one of the Program Committee Members of the International conference on Database Management Systems (DMS-2019).

Introduction

This book is an essential guide to the implementation of image processing techniques, with tutorial introductions and sample code in MATLAB. Algorithms are presented and fully explained to enable complete understanding of the methods and techniques demonstrated. The main strength of the proposed book is the exemplar code of the algorithms written in Matlab. Fully updated with the latest developments in contour extraction & compression, including expanded tutorials and new techniques, this book contains extensive new material on DCT, Haar & wavelets.

The principal objectives of the book continue to be to provide an introduction to basic concepts and methodologies for digital contour approximation, and to develop a foundation that can be used as the basis for further study and research in this field. The material is timely, highly readable, and illustrated with numerous examples of practical significance. Image and contour compression mainstream areas of image processing in the spatial and frequency domains, are covered. Coverage concludes with a discussion on the fundamentals of object recognition. Chapters supported by figures, examples, illustrative problems, and exercises. Useful websites and an extensive list of bibliographical references. This accessible text is ideal for upper-level undergraduate and graduate students in digital image and video processing courses, as well as for engineers, researchers, software developers, practitioners, and anyone who wishes to learn about these increasingly popular topics on their own.

This is an unusual book. It is a book of exercises, exercises in digital imaging engineering, one of the most important and rapidly developing branches of modern information technology. Studying digital imaging engineering, mastering this profession and working in the area is not possible without obtaining practical skills based on fundamental knowledge in the subject. The current book is aimed at providing technical support for this. It contains exercises on all major topics of digital imaging for students, researchers in experimental sciences and, generally, all practitioners in imaging engineering.

Chapter One

Matlab Basics Operations

Matlab is a programming tools and environment deals with arithmetic purposes, which it has a lot of jobs and mathematics functions which are built internally and that is easiest to solve a different types of mathematics equations. It helps this programming language to write special functions and programs. In addition to the lot of the other characteristics. Some books that are simple such that you aren't lost from the start because the text assumes you are very experienced with MATLAB and/or image processing, and another which is almost at that level to use as a reference guide [1] and [2].

The Matlab can uses in the following fields:

- Mathematics and computation
- Algorithm development
- Data acquisition
- Modeling, simulation, and prototyping
- Data analysis, exploration, and visualization
- Scientific and engineering graphics
- Application development, including graphical user interface building

For more information about Matlab please visit the following company site that producing for the system:
http://www.mathworks.com/

1.1 Special Characters

% [] : ; , ()

```
>> % percent symbol denotes a comment.

>> R = [1   4   2   6] % brackets are used to form vectors and matrices.
R =
   1  4  2  6

>> T = 1 : 4 % (colon is used to create vectors, matrix subscripting, and for iterations).
T =
   1   2   3   4

>> D = 0  :  0.1  :  0.5  % using two colons starts from 0 to 0.5 with step 0.1
D =
   0  0.1  0.2  0.3  0.4  0.5
```

>> S = [16 2 3 13; 5 11 10 8; 9 7 6 12; 4 14 15 1] % semicolon is used inside brackets to end rows and also used to suppress printing or separate statements.

```
S =
   16    2    3   13
    5   11   10    8
    9    7    6   12
    4   14   15    1
```

Hint: [ ] is used also to store empty matrix as in the following example:

>> S(2 : 3, :) = [ ] % eliminate the elements in the second and third rows and the all columns.

```
S =
    16    2    3   13

     4   14   15    1
```

>> S(3,4) % (coma is used to separate matrix subscripts and function arguments; where 3 and 4 is the number of rows and columns respectively).

```
ans =
   12
```

>> [S(1,1) S(2,3) S(3,4)] % parentheses are used to indicate precedence in arithmetic expressions in the usual way.

```
ans =
    16   10   12
```

1.2 Arithmetic Operators

‘ + - * .* \ .\ / ./ ^

```
>> A= [5    3; 0    2]
A =
   5   3
   0   2
```

>> A’ % matrix transpose

```
ans =
   5   0
   3   2
```

```
>> B= [4    6; 1    7]
B =

   4   6
```

```
    1    7

>> A + B  % addition
ans =
    9    9
    1    9

>> A – B  % subtraction

ans =
    1   -3
   -1   -5

>> A * B  % linear algebraic product
ans =
    23    51
     2    14

>> A.* B  % array multiplication (element by element product)
ans =
    20    18
     0    14

>> A \ B  % matrix left division
ans =
    0.5000   -0.9000
    0.5000    3.5000

>> A. \ B  % array left division (is the matrix with elements B(i,j) / A(i,j)
ans =
  0.8000    2.0000
     Inf      3.5000

>> A / B  % matrix right division (is the same as A * inv(B); where inv is the inverse of the matrix)

ans =

    1.4545   -0.8182

   -0.0909    0.3636


>> A. / B  % array right division (is the matrix with elements A(i,j) / B(i,j)

ans =
```

```
    1.2500    0.5000
         0    0.2857
>> A ^ 2  % matrix power
ans =
    25    21
     0     4

>> A. ^ B  %  array power
ans =
   625   729
     0   128

>> AB= [5  33   3; 0  47   132]

AB =

     5    33     3
     0    47   132

>> size(AB) % compute matrix size

ans =

     2     3
```

Means that the matrix has 2 rows & 3 columns

```
>> sum(sum(AB)) % calculate the summation of all elements in the matrix

ans =

   220
```

Hint: in case of the vector we need only to write sum(AB)

```
>> H=[1 2;3 4]

H =

   1   2
   3   4

>> F=find(H<4)

F =

   1
   2
   3
```

1.3 Relational operations

> < >= <= == ~= ~

Elements set to one where the relation is true, and elements set to zero where it is not.

```
>> A > 2
ans =
   1   1
   0   0

>> A ~= 3
ans =
   1   0
   1   1
```

1.4 Logical operators

& | ~ is the logical operators AND, OR, and NOT.

```
>> x = [1   0; 1   1]
ans =
    1   0
    1   1
>> y = [1   0; 0   0]
ans =
    1   0

    0   0
>> x & y
ans =
    1   0
    0   0
>> x  |  y
ans =
    1   0
    1   1
>> ~ x
ans =
    0   1
    0   0

>> x=[1 3 -15 -2 9]

x =

    1   3  -15   -2   9

>> z=[1 0 0 0 -2]
```

```
z =

    1    0    0    0   -2

>> mu= conv (x,z)  % multiply two polynomial equations in each other

mu =

  Columns 1 through 9

    1    3  -15   -2    7   -6   30    4  -18
>> [d, r]= deconv (mu, x) % % divide two polynomial equations on each other where d is division product vector & r is the division residues vector


d =

    1    0    0    0   -2



r =

  Columns 1 through 9

    0    0    0    0    0    0    0    0    0

>> polyval (x,3) % calculate values of polynomial value at a certain value 3

ans =

   30
```

1.5 Commands and functions

```
>> Y = -12 +  4i   % i is the imaginary part
Y =
     -12.0000 + 4.0000i
>> abs( Y )  % absolute value and string to numeric conversion
ans =
    12.6491
>> x = -1   :   0.5   :   1;
>> plot (x, asin(x))  % plot the three circular inverse trig functions for the range [-1, 1]
```

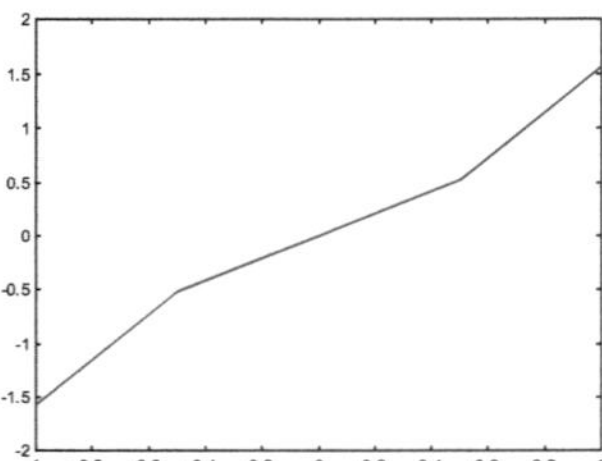

Fig. 1.1. Color image with size (3 x 3)

>> axes('position',[0.1   0.1   0.8   0.6]); bar (peaks); axes('position',[0.1   0.7   0.8   0.2]); pcolor ([1 : 10; 1 : 10]) % plot multiple axes in a single figure window where 0.1, 0.1, 0.8 and 0.6 is the left , bottom, width and height of the rectangle respectively.

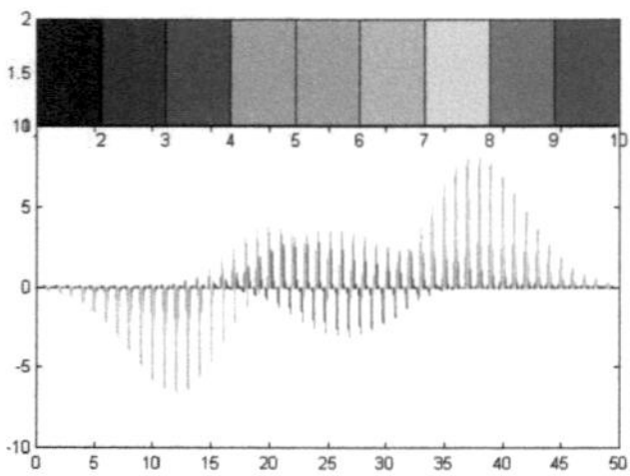

Fig. 1.2. Color image with size (3 x 3)

>> x = 0 : 0.01 : pi/2; plot (x, tan(x))

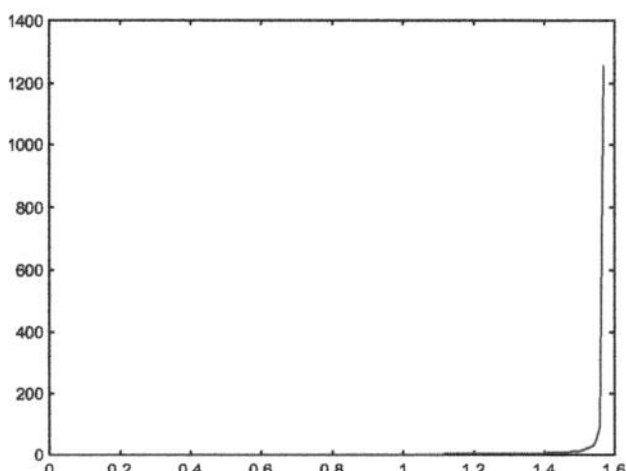

Fig. 1.3. Color image with size (3 x 3)

but if the last statements is followed by

>> axis ([0 pi/2 0 10]) % axis scaling and appearance in a plot where 0, pi/2, 0 and 10 are the xmin, xmax, ymin and ymax respectively; then a much more satisfactory Plot results.

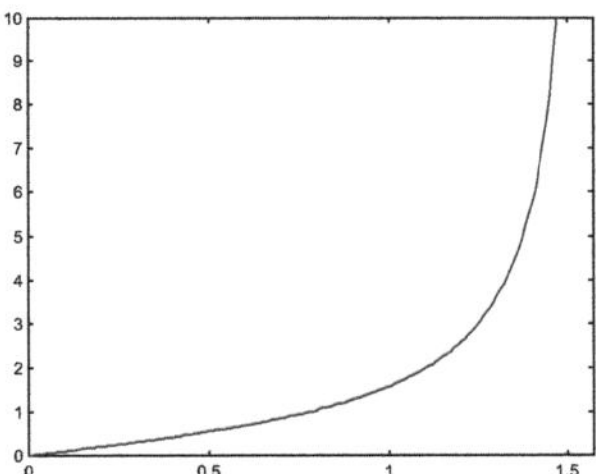

Fig. 1.4. Color image with size (3 x 3)

>> S = zeros(4,4) % create matrix with 4 rows and 4 column with zeros elements.

0 0 0 0

```
   0   0   0   0
   0   0   0   0
   0   0   0   0

>> S = ones(4,4)  % create matrix with 4 rows and 4 column with ones elements.
   1   1   1   1
   1   1   1   1
   1   1   1   1
   1   1   1   1
>> isempty(S)  % detect empty matrix and return 1 if S is an empty matrix and 0 otherwise.
   0
>> eye(3)  % is the 3 x 3  Identity matrix with 1's on the diagonal and zeros elsewhere.
   1   0   0

   0   1   0
   0   0   1
>> Z = [10  4  7  9; 54  82  15  0; 61  92 32  45;36  87  26  19]
Z =
    10    4    7    9
    54   82   15    0
    61   92   32   45
    36   87   26   19
>> size(Z)  % determine the matrix size
ans =
    4    4
>> max(max(Z))  % maximum number in the matrix
ans =
    92
>> min(min(Z))  % minimum number in the matrix
ans =
   0
>> diag(Z)  % diagonal of the matrix
```

```
ans =
   10
   82
   32
   19
>> diff(Z)  % is the matrix of column differences.
ans =
    44   78    8   -9
     7   10   17   45
   -25   -5   -6  -26
>> tril(Z)  % extract lower triangular part

ans =
    10    0    0    0
    54   82    0    0
    61   92   32    0
    36   87   26   19
>> triu(Z)  % extract upper triangular part
ans =
    10    4    7    9
     0   82   15    0
     0    0   32   45
     0    0    0   19
>> flipdim(Z,1)  % opposite the rows
ans =
   36   87   26   19
   61   92   32   45
   54   82   15    0
   10    4    7    9
>> flipdim(Z,2)  % opposite the columns
ans =
    9    7    4   10
```

```
  0  15  82  54
 45  32  92  61
 19  26  87  36
>> flipud(Z)  % flip matrix in up/down direction
ans =
    36  87  26  19
    61  92  32  45
    54  82  15   0
    10   4   7   9
>> fliplr(Z)  % flip matrix in left/right direction


ans =
     9   7   4  10
     0  15  82  54
    45  32  92  61
    19  26  87  36

>> rot90(Z)  % rotate matrix 90 degrees
ans =
     9   0  45  19
     7  15  32  26
     4  82  92  87
    10  54  61  36
>> mean(mean(Z))  % average or mean value of the matrix
ans =
   36.1875
>> median(mean(Z)) % median value  of the matrix
ans =
   32.7500
>> [I ,J] = find(Z>45)  % returns the row(I) and column (J) indices of the nonzero entries
in the matrix Z
I =
```

```
    2
    3
    2
    3
    4
J =

    1
    1

    2
    2
    2
>> H = [17   33   6   0   99   72   8]
H =
     17   33   6   0   99   72   8
>> sort ( H )  % sort in ascending order
ans =
      0   6   8   17   33   72   99
>> a =2.35
a =
     2.3500
>> floor (a)  % round towards minus infinity
ans =
     2
>> ceil (a)  % round towards plus infinity
ans =
     3
>> round (a)  % round the elements of a towards nearest integer
ans =
     2
```

1.6 Important Statements

- if

If if statement condition.

The general form of the IF statement is :

```
if expression
   statements
elseif expression
   statements
end
```

- for

for Repeat statements a specific number of times.

The general form of a for statement is:

```
for variable = expression
    statement
      ...
    statement
 end
```

Example 1.1:

```
for i = 1 : 4
   for j = 1 : 4
      if i = = j
         a(i,j) = 2;
      elseif abs([i,j]) = = 1
         a(i,j) = 1;
       else
         a(i,j) = 0;
       end
    end
end
```

- while

while repeat statements an indefinite number of times.

The general form of a while statement is:

```
while expression
    statements
end
```

Example 1.2:

```
A = [0.1963  -0.3927  0.5890  -0.7854  0.9817  -1.1781  1.3744  -1.3744];
i = 1;slope = [ ];
while (isempty(A(i))= = 0),
   slop = tan(A(i));
   i = i+1;
   slope = [slope slop];
   plot(slope)
end
```

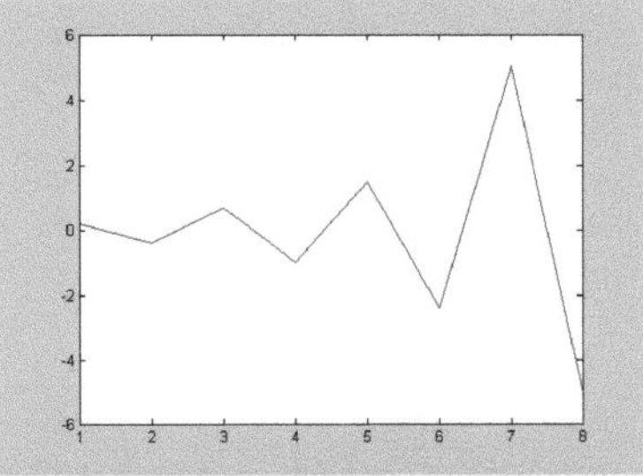

Fig. 1.5. Color image with size (3 x 3)

Example 1.3: Draw y=sin(x) where $-\pi \leq x < \pi$ with step 0.1. Put x as a title in x axis and $y = \sin(x)$ as a title in y axis using Matlab ?

Solution:

>> x=-pi:0.1:pi;y=sin(x); % calculate the sinusoidal function in range of x is $[-\pi,\pi]$

>> plot (x,y) % xy plot

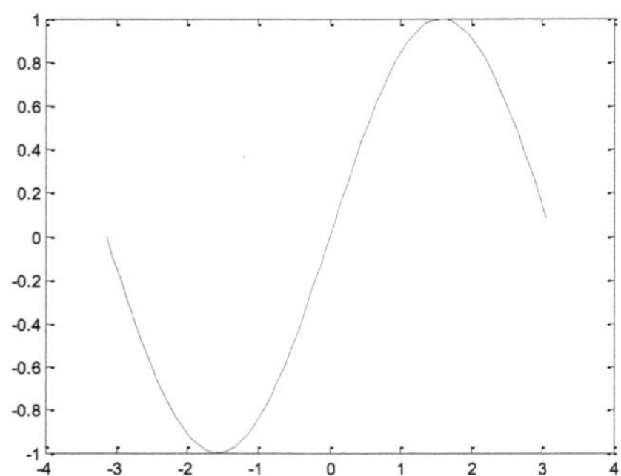

>> plot(x,y,'s-') % mark the points with square

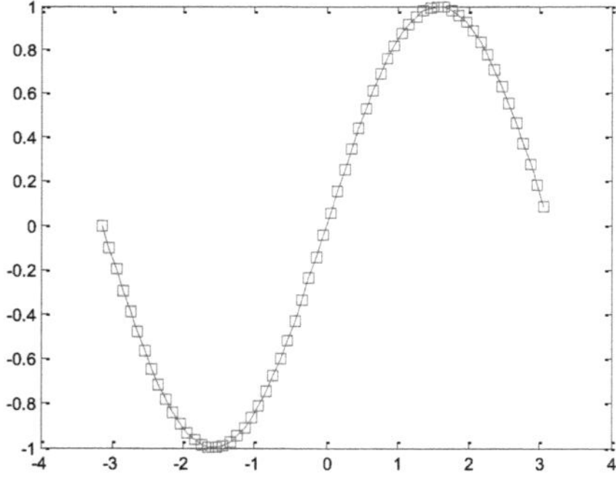

>> xlabel('x');ylabel('y=sin(x)') % labels the the plot as 'x' and 'y=sin(x)'

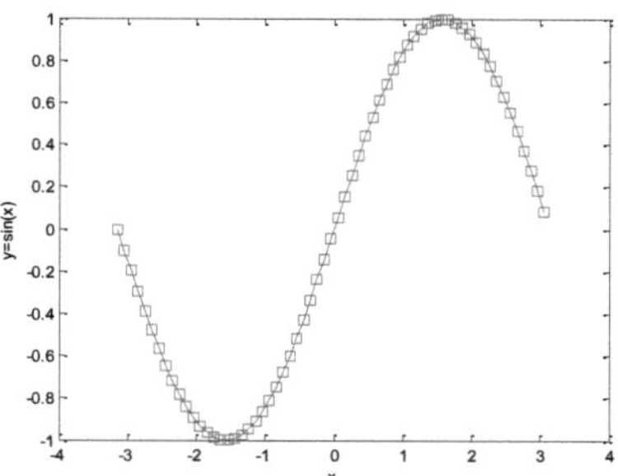

1.7 Image Formats

```
>> f = [-0.5  0.5; 0.75  1.5]  % f is image (matrix) in double format

f =

   -0.5000    0.5000
    0.7500    1.5000

>> g = im2uint8(f)  % f is converted to uint8 format

g =

     0  128
   191  255

>> h = im2double(g)  % g is converted to double format scaled to range [0,1]
 h =

        0    0.5020
   0.7490    1.0000
```

```
>> x = [25  50; 128  200]  % x is intensity image
x =

   25   50
  128  200


>> d= im2bw(x)  % intensity image is converted to black & white (logical) image

d =

   1   1
   1   1

>> e= mat2gray(f)  % convert f in double format to the range [0, 1]
e =

        0    0.5000
   0.6250    1.0000

>> i= im2bw(e,0.1)  %  now convert e in range [0, 1] to the binary image using threshold 0.1
i =

   0   1
   1   1
```

1.8. Image Types

There are four main important images types stored by Matlab:

1) Intensity Images: the image has m-by-n matrix for integer (uint8 and uint16) values and non-integer values (double) which scaled and linear to obtained color

map indices. The values general range is [0, 1] for double data and [0, 255] or [0, 65535] for uint8 and uint16 respectively. Color map is matrix p-by-3 for non-integer values in the range [0, 1].

>> I=imread ('rice.tif'); % display the gray level (intensity) image store in an uint8 format (Fig. 1.6)

Fig. 1.6. Intensity Image (Tools)

```
>> whos I
  Name      Size         Bytes  Class    Attributes

  I       256x256        65536  uint8
```

Grand total is 65536 elements using 65536 bytes

I is an array of class uint8, and size 256x256. That means 256 rows and 256 columns. We can see some of this information with the following commands

```
>> class(I)
ans =

uint8
```

We will talk later on about the different image classes.

Sometimes it is useful to determine the number of rows and columns in an image. We can achieve this by means of

```
>> [M, N] = size(I);
```

To display the image we use imshow

>> imtool(I)

In the figure window we have now available the following tools: overview, pixel region, image information, adjust contrast and zoom. Try them.

Images can be written to disk using the function imwrite. Its format is

imwrite(f, 'filename')

with this syntax, filename must include the file format extension. Alternatively

imwrite(f, 'filename', format)

saves I using format. For example

>> imwrite(I, 'test', 'jpeg', 'quality', 25)

In the help you can find more information about available formats and their options.

>> imtool (I) % we may see the pixel values (from 0 to 255) of the image by moving the cruse pointer like

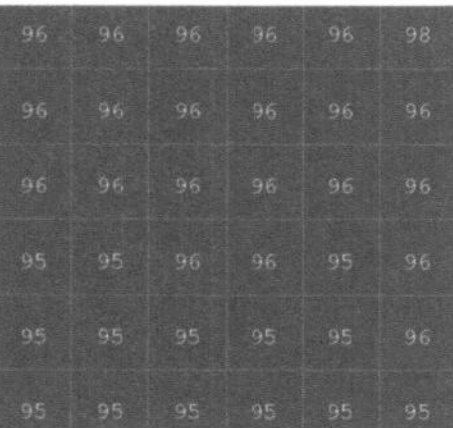

Fig. 1.7. Pixels displayed

- Contrast Adjustments

Often, images have a low dynamic range and many of its features are difficult to see. We will present different intensity transformations that will improve the appearance of the images. Improving the appearance of an image does not merely serve an aesthetic role – often, it can help improve the performance of image segmentation algorithms and feature recognition.

During contrast adjustment, the intensity value of each pixel in the raw image is transformed using a transfer function to form a contrast-adjusted image. The most common transfer function is the gamma contrast adjustment:

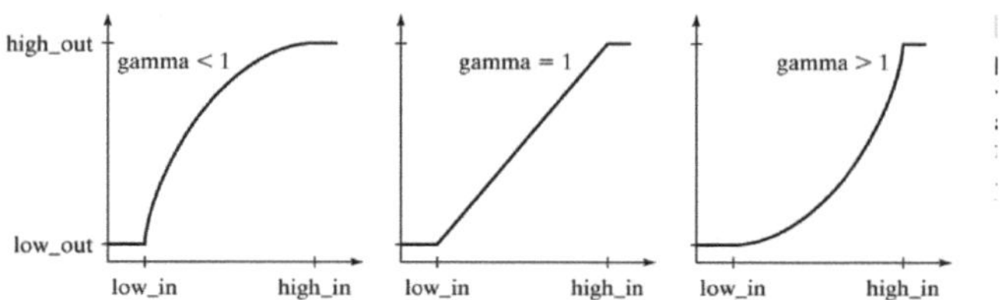

Here low_in and low_high give the low and high grayscale intensity values for the contrast adjustment, and gamma gives the exponent for the transfer function.

The basic matlab function for contrast adjustment is imadjust. It has the syntax:

g = imadjust(I,[low_in high_in],[low_out high_out],gamma)

Without the optional arguments (g = imadjust(I)), the output image has 1% of the data in I is saturated at low and high intensities.
Try the following commands (negative image)

```
>> f = imread('tools.bmp');

>> imshow(I)

>> g1 = imadjust(I);

>> imshow(g1)
```

```
>> g2 = imadjust(I, [0 1], [1 0]);

>> figure,imshow(g2)
```

>> imagesc(I,[100 222]); colormap(gray) % Scale data (from minimum value 100 to maximum value 222) and display as image (see Fig. 1.8)

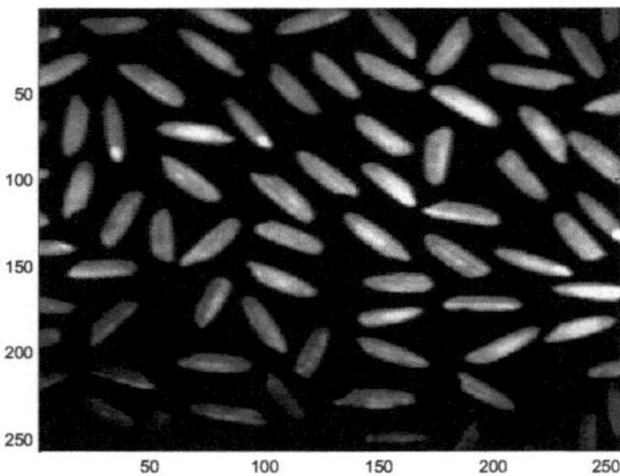

Fig. 1.8. Scaled image (100-222)

2) RGB Images: the image has m-by-n-by3 with non-integer values in the range is [0, 1] for double data and with integer values in the range [0, 255] or [0, 65535] for uint8 and uint16 respectively.

>> RGB_Color= imread ('lena.jpg'); % read the color image ('Lena') stored as JPG format

>> figure; imshow (RGB_Color, []); % display the color image ('Lena') % (see Fig. 1.9)

Fig. 1.9. Color image (Lena)

>> R= RGB_Color (:,:,1);figure; imshow (R,[]) % extracted and display red component image of color image

Fig. 1.10. Red component of the Color image

>> G= RGB_Color (:,:,2); figure; imshow (G,[]) % extracted and display green component image of color image

Fig. 1.11. Green component of the Color image

>> B= RGB_Color (:,:,3); figure; imshow (B,[]) % extracted and display blue component image of color image

Fig. 1.12. Blue component of the Color image

3) Indexed Images: the image has m-by-n matrix in the range is [1, p] for double data and [0, p-1] for uint8 and uint16 respectively, where p represented color map which is matrix p-by-3 from non-integer values in the range [0, 1].

>> [X, map]=gray2ind(I,100); % convert from gay level image to indexed image where 100 represented the colors number in the image
>> figure; imshow (X, map) % display the indexed image included it's color map

Fig. 1.13. Indexed image with color map for intensity image

If the image is color the we can use
>> [X,map]=rgb2ind(RGB_Color,100); % convert from color image to indexed image where 100 represented the colors number in the image
>> figure; imshow (X, map) % display the indexed image included it's color map

Fig. 1.14. Indexed image with color map for color image

4) Binary Images: the image has m-by-n matrix in the range is [1, p] for double data and [0, p-1] for uint8 and uint16 respectively, where p represented color map which is matrix p-by-3 from non-integer values in the range [0,1].

We can calculate the threshold value automatically by using
>> level = graythresh (image);
Now create a binary image using the obtained threshold value:

>> imb = im2bw (image, level);
Let us concern the following example
>> I= imread ('rice.tif');
>> BWG= im2bw (X, map); figure; imshow (BWG,[]) % convert intensity image to binary image the displayed it

Fig. 1.15. Binary image of intensity image

>> BWC=im2bw (RGB_Color); % convert color image to binary image

>> figure; imshow (BWC) % display the binary image of color image

Fig. 1.16. Binary image of color image

Chapter two
Contour Extraction Explanation

2.1 Introduction

There exist two main streams of research for the 2D-object contour extraction problem: The Object Contour Following (OCF) or sequential methods and the Multiple Step Contour Extraction (MSCE) or parallel methods. Later, they have developed an Object-oriented Contour Extraction algorithm (OCE). Also there exist algorithm was initiated a third class of contour Extraction algorithms, Single Step Parallel Contour Extraction (SSPCE). In this explanation we follow the steps which should be done for the grey-level 2-D image to obtained the contour extracted in y-x Cartesian co-ordinates representation. These steps are shown also using Matlab programming.

2.2 Color & Binary Images

Figure 2.1 shows a general block diagram (the steps) of the contour extraction process.

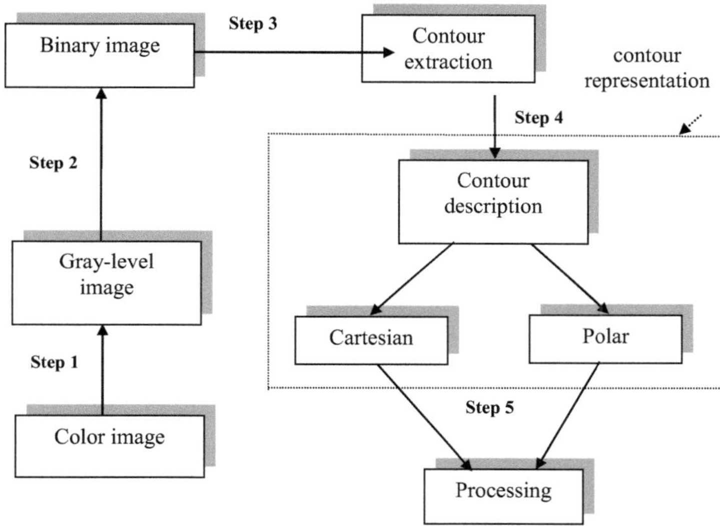

Fig 2.1. Block diagram of contour extraction

To begin with digital image is delivered (usually on CD) as a raster file. A raster image file is generally defined to be a rectangular array of regularly sampled values, known as pixels. By definition a raster file is simply a file organized in rows and columns of values. For a digital image each row/column address has either a grey-scale or color value. Each pixel (picture element) has one or more numbers associated with it, generally specifying a color which the pixel should be displayed in. The term pixel depth comes in here, you may have heard the terms 8 bit, 24 bit, palliated, and true color. These terms just refer to what information is stored at each row/column or pixel address. Fig. 2.2 shows a super-simple image definition (3x3 image from upper left, black, medium red, full red, medium grey, medium green, full green, white, medium blue, full blue) and the related values of pixels are shown in Table 2.1.

Fig. 2.2. Color image example (3 x 3 pixels)

Table 2.1

Red	Green	Blue	Red	Green	Blue	Red	Green	Blue
0	0	0	128	0	0	255	0	0
128	128	128	0	128	0	0	255	0
255	255	255	0	0	128	0	0	255

Each small square you see is an individual pixel in the original image. Pixel is a computer term for "picture element". Each square in this grid is a pixel. All images are always rectangular, regardless if the back ground pixels have been made transparent or not (as here). This, however, is a very simple way of representing an image and takes a very large amount of disk space. Most image formats, therefore, are compressed to shrink the size of the image. Additionally, some image formats allow for a much smaller range of colors (example: GIF is an 8 bit format and allows for only 256 different colors to be present in any image).

Image formats such as this also, generally, do not directly specify colors in the manner shown above. Instead, each pixel is represented by a single number (for 8 bit images, a value from 0-255) which is an index into a color lookup table. Thus, each pixel only requires 8 bits (1/3 of the space) to represent it. The color lookup table then specifies colors in the full 24 bit true-color representation but it is so small relative to a normal sized image that the space it takes up is negligible.

As a conclusion we can see a digital color image pixel is just a RGB data value (Red, Green, and Blue). Each pixel's color sample has three numerical RGB components (Red, Green, and Blue) to represent the color. These three RGB components are three 8-bit numbers for each pixel. Three 8-bit bytes (one byte for each of RGB) is called 24 bit color. Each 8 bit RGB component can have 256 possible values, ranging from 0 to 255.

In the base 2 binary system, an 8 bit byte can contain one of 256 numeric values ranging from 0 to 255, because 2 to the 8th power is 256, as seen in the sequence 2,4,8,16,32,64,128,256. The 8th of these is 256. For example, three values like (250, 165, 0), meaning (Red=250, Green=165, Blue=0) to denote one Orange pixel.

Photo editor programs have an Eye Dropper tool to show the 3 RGB color components for any image pixel.

To follow the steps as in Figure 2.1, let us take the following example.

1. Step 1 (from the color image to the grey-scale image).

Suppose we have color image (back arrow) as shown in Figure 2.3 with size of (22 x 22) pixels.

To call or read the color image using Matlab programming we wrote:

Fig. 2.3. The color image (back arrow)

>> A = imread ('backarrow.jpeg'); % color image store it by jpeg format.

To define the size of the image we can write

>> [n, m] = size(A); % we define the size of the color image where n and m are the number of rows and columns respectively.

To show this color image in a figure we write

>> figure; imshow (A, [ ]);

Next we want to transfer this color image to the grey-level image by writing

>> I = rgb2gray(A); %

To show this grey-level image in a figure we can write

>> figure; imshow (I, [ ]);

The grey-level image is shown in Fig. 2.4.

Fig. 2.4. The grey-scale image (back arrow)

The pixels values of the matrix for this image (I_{ij}) are shown in Table2.2 and Table2.3 where *i* and *j* are the row and column number of the matrix respectively.

Table 2.2

i \ *j*	1	2	3	4	5	6	7	8	9	10	11
1	250	250	252	252	252	251	250	250	250	253	249
2	251	252	252	253	253	252	251	252	253	254	246
3	254	253	254	254	254	254	254	254	253	254	253
4	253	254	255	255	254	253	253	253	247	252	250
5	252	253	253	254	253	252	252	253	250	249	233
6	253	253	253	253	255	253	252	253	250	241	202
7	254	254	253	253	253	254	253	250	236	212	173
8	253	254	253	253	252	251	249	245	195	175	157
9	253	252	254	253	248	248	238	199	166	173	168
10	252	251	250	254	251	237	198	167	158	166	165
11	250	254	253	252	246	199	156	167	171	170	170

Table 2.3

i \ *j*	12	13	14	15	16	17	18	19	20	21	22
12	167	163	167	168	167	214	250	252	253	253	251
13	169	158	160	164	155	212	248	253	253	253	252
14	169	161	170	178	162	219	249	250	253	253	252
15	164	174	208	235	224	233	252	250	252	253	252
16	163	175	222	249	251	247	253	252	251	253	253
17	167	170	225	249	248	253	254	254	254	254	254
18	206	189	221	250	251	253	255	255	255	255	255
19	243	220	230	249	251	253	255	255	255	255	255
20	252	245	247	251	250	254	255	255	255	255	255
21	253	252	252	252	251	254	255	255	255	255	255
22	253	250	248	253	253	254	255	255	255	255	255

2. Step 2 (from the grey-level to the binary image).

This step is to convert the grey-level image to the binary image (i.e. the values of pixels has only zeros or ones and no more). To do that the edge detection is required as the first step to simplify images and to facilitate image analysis and interpretation. As we can see from the Tables (2.2 and 2.3) the range of the value of the grey-level is from 155 to 255.

The idea is like that: we want to put the values of the pixels which are smaller than the threshold (let us say 190) is ones and the others are zeros.

Using Matlab programming we can write

>> D = B < 190; % the pixels which has values smaller than 190 we will put it ones and that pixels grater than 190 will be zeros.

To show this in a figure we write

>> figure; imshow(D, [ ]);

The binary image (D) is shown in Figure 2.5.

Fig. 2.5. The binary image (back arrow)

The matrix of this image is shown in the Table 2.4. The idea is to fill the image with white color (i.e. ones) and the outside with black color (i.e. zeros). Finally we obtained the contour of the back arrow image as we can see it from the Table 2.4.

3. Step3 (contour extraction from the binary image).

In this step we will show how the contour as shown in Fig. 2.5 to be extracted. Detection of edge points (pixels) of a 3-dimentional physical object in a 2-dimensional image and contour is one of the main research areas of computer vision. Edge detection is required as the first step to simplify images and to facilitate image analysis and interpretation. Edge detection extracts and localizes points (pixels) around which a large change in image brightness has occurred. Edge detection is based on the relationship a pixel has

with its neighbors. If the grey-levels around a pixel are similar, the pixel is unsuitable to be recorded as an edge point. Otherwise, the pixel may represent an edge point.
There exist methods for contour extraction based on the window (window, 3 x 3 window and 4 x 4 window). In our examples we used the algorithms is known as Single Step Parallel Contour Extraction 'SSPCE', uses a 3x3 window and will be described later in more details.

Table 2.4 (the binary image)

0	0	0	0	0	0	0	0	0	0	0	0	0	0	0	0	0	0	0	0	0	0
0	0	0	0	0	0	0	0	0	0	0	0	0	0	0	0	0	0	0	0	0	0
0	0	0	0	0	0	0	0	0	0	0	0	0	0	0	0	0	0	0	0	0	0
0	0	0	0	0	0	0	0	0	0	0	0	0	0	0	0	0	0	0	0	0	0
0	0	0	0	0	0	0	0	0	0	0	0	1	0	0	0	0	0	0	0	0	0
0	0	0	0	0	0	0	0	0	0	0	1	1	0	0	0	0	0	0	0	0	0
0	0	0	0	0	0	0	0	0	0	1	1	1	0	0	0	0	0	0	0	0	0
0	0	0	0	0	0	0	0	0	1	1	1	1	0	0	0	0	0	0	0	0	0
0	0	0	0	0	0	0	0	1	1	1	1	1	1	1	1	0	0	0	0	0	0
0	0	0	0	0	0	0	1	1	1	1	1	1	1	1	1	0	0	0	0	0	0
0	0	0	0	0	0	1	1	1	1	1	1	1	1	1	1	0	0	0	0	0	0
0	0	0	0	0	0	1	1	1	1	1	1	1	1	1	1	0	0	0	0	0	0
0	0	0	0	0	0	0	1	1	1	1	1	1	1	1	1	0	0	0	0	0	0
0	0	0	0	0	0	0	0	1	1	1	1	1	1	1	1	0	0	0	0	0	0
0	0	0	0	0	0	0	0	0	1	1	1	1	0	0	0	0	0	0	0	0	0
0	0	0	0	0	0	0	0	0	0	1	1	1	0	0	0	0	0	0	0	0	0
0	0	0	0	0	0	0	0	0	0	0	1	1	0	0	0	0	0	0	0	0	0
0	0	0	0	0	0	0	0	0	0	0	0	1	0	0	0	0	0	0	0	0	0
0	0	0	0	0	0	0	0	0	0	0	0	0	0	0	0	0	0	0	0	0	0
0	0		0	0	0	0	0	0	0	0	0	0	0	0	0	0	0	0	0	0	0
0	0	0	0	0	0	0	0	0	0	0	0	0	0	0	0	0	0	0	0	0	0
0	0	0	0	0	0	0	0	0	0	0	0	0	0	0	0	0	0	0	0	0	0

2.3 Contour extraction methods

- **Freeman chain coding:**

The codes associated with four or eight possible directions are the chain codes and, with as the current contour pixel position.

Freeman chain coding can be used to distinguish all possible connections for both 8-connectivity and 4-connectivity schemes. An 8-directional chain coding uses eight possible directions to present all possible line segments connecting nearest neighbors according to the 8-connectivity scheme as shown in Figure 2.6.

We follow the contour in a clockwise manner and keep track of the directions as we go from one contour pixel to the next. For the standard implementation of the chain code we consider a contour pixel to be an object pixel that has a background (non-object) pixel as one or more of its 4-connected neighbors.

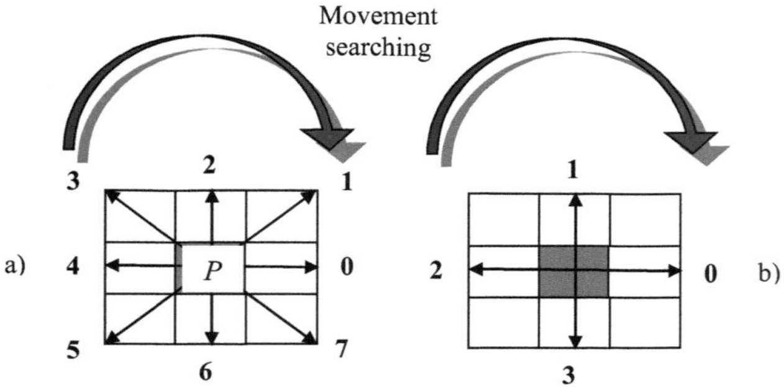

Fig. 2.6. Freeman chain coding. a) 8-Directional, b) 4-Directional

Chain code properties:

- Even codes {0,2,4,6} correspond to horizontal and vertical directions; odd codes {1,3,5,7} correspond to the diagonal directions.
- Each code can be considered as the angular direction, in multiples of 45^0, that we must move to go from one contour pixel to the next.
- The absolute coordinates $[m,n]$ of the first contour pixel (e.g. top, leftmost) together with the chain code of the contour represent a complete description of the discrete region contour.

- When there is a change between two consecutive chain codes, then the contour has changed direction. --This point is defined as a *corner*.

The image is scanned from left to right and from top to bottom. All stored pixels are marked for avoiding multiple coding.

The contour extraction from 2D-image plays an important role in image extraction, recognition and analysis.

Contour extraction is frequently used in visual information processing [3], [4],[5], and [6]. The contour of an object contains relevant information about the shape of that object [7]. Shape Analysis and Retrieval of Multimedia Objects is designed to meet the needs of practitioners and researchers in industry, and graduate-level students in Computer Science [8].

Hence, object contour extraction from given 2D-digital images is an important step for feature-based object recognition. There exist two main streams of research for contour extraction of a 3D-object in a 2D- image [9]:

1. Object Contour Following (OCF) which is classified as sequential methods [32] and [33], and
2. Multiple Step Contour Extraction (MSCE) which is classified as parallel methods [9].

Recently, an Object-oriented Contour Extraction 'OCE' has been developed which uses 4x4 windows [9] and [10].

By sequential methods, it is meant that the result at a point is dependent upon the result of the previously processed points. By parallel, it is meant that the decision of whether or not a point is on an edge is made on the basis of the gray level of the point and its neighbors. So the edge detection operator in principle can be applied simultaneously everywhere in the picture.

Another class of contour extraction algorithms is known as Single Step Parallel Contour Extraction 'SSPCE' [11] and [12], which uses a 3x3 window, and will be described in this chapter. A new algorithm which uses 2x2 windows [12] was proposed in order to be simpler than the previous ones and more suitable for real time applications. Some books were written in order to develop a generative theory of shape that has two properties regarded as fundamental to intelligence - maximizing transfer of structure and maximizing recoverability of the generative operations [13] and [14].

2.3.1 Object contour following 'OCF' method

This method is used to trace the contour edges of a 2-D digital image. This algorithm uses a 4x4 pixel window structure to extract the object contours. The extraction procedure consists of finding a starting point and then crossing the edge between the white and black regions, recording the co-ordinates of the black pixel then turning left continuously until a white pixel is found, recording the black pixel co-ordinates as the next contour edge point. The begin turning right until a black pixel is found. Terminate this procedure when the starting point of the contour is reached again. The major problems are that picking a proper starting point, which is not always an easy task (in some cases an additional procedure for selecting an appropriate starting point is required) and a termination criterion is required to determine when the procedure is finished (in some cases the procedure should be terminated if it is unable to reach the starting point). The idea of these methods is illustrated in Fig. 2.7.

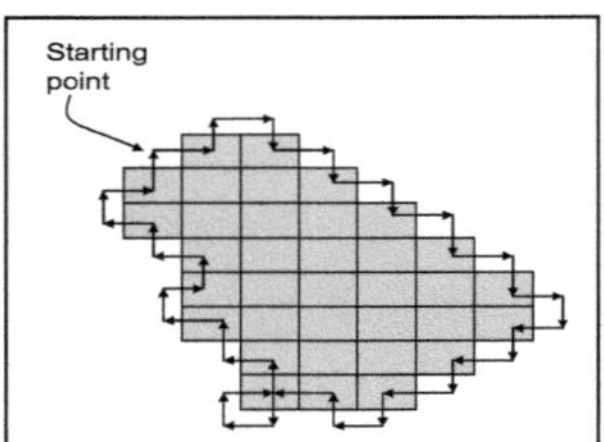

Fig. 2.7. Object contour follower (OCF)

OCF algorithms have three main drawbacks [9]:

1. Possible extraction of a false object contour starting point,
2. Possible extraction of a false contour (implement contour), and
3. Long computation time whenever several contours are to be extracted.

The main problem arising from extracting object contours in real applications is the effect of noise [15]. Whenever the original 2D- digital image is noisy, the effect of the first two drawbacks leads to incomplete contours [16]. The third drawback is a result of the sequential nature of the OCF algorithms.

2.3.2 Multiple step contour extraction 'MSCE' method

The key idea is to transform the original grey 2D-digital image stepwise such that only object contour points remain marked [9]. The algorithmic steps are given below.

In using the MSCE algorithms, there are essentially three steps [9]:

Detection of contour points, upgrading the contour image, and chain coding contour points. Fig. 2.8 shows the steps required for extraction object contours by these methods.

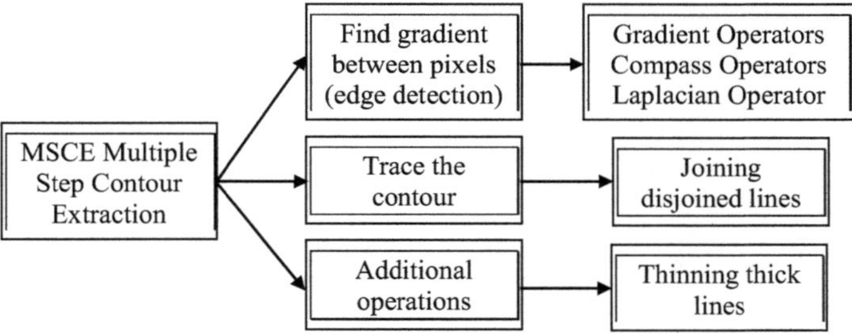

Fig. 2.8. Block diagram of the Multiple Step Contour Extraction (MSCE) method

The difference between two pixels is the gradient between them with respect to different gray scale levels, and the gradient will be zero for the pixels with the same gray level. An additional procedure is used to improve the overall contour structure by joining all disjoined edges and thinning the thick edges. This algorithm uses a 3x3 pixels windows structures to extract the object contours. Although the method of gradient operator for generating edge elements is parallel, the method of connecting (tracing) these extracted edge elements is sequential. Like the OCF algorithms, the quality of contour extraction is influenced by the image noise. Depending on the image noise, diverse problems may arise such as smeared contour points, discontinuous contours, and pseudo contour points [9].

2.3.3 Object-oriented contour extraction 'OCE' method (4x4 windows)

This algorithm uses a 4x4 pixels window structure to extract the object contours by the four central pixels which are processed simultaneously. The algorithm uses the features of both OCF and MSCE methods to overcome most of the disadvantages they have. It features a parallel implement and an effective suppression of noises. It can be realized in real-time [9]. The following three steps are needed for the extraction procedure:

Step1: The image is framed with zeros.

Step2: Eight rules of edge extraction are applied and are coded using 8-directional chain-code as shown in Listing 2.1.

Listing 2.1
Implementation of the eight rules for contour extraction (4x4 windows)

```
a(i,j) ←0; i = 1,2,.....,N; j = 1,2,.....,N;
for i = 2,3,.....,N-1; j = 2,3,.....,N-1;
{
if b(i,j+1) and b(i+1,j+1) and [b(i,j+2) or b(i+1,j+2)]
then a(i,j+1) ←a(i,j+1) or 2^0                    { edge 0 }
if b(i+1,j) and b(i,j+1) and b(i+1,j+1)
then a(i,j+1) ←a(i,j+1) or 2^1                    { edge 1 }
if b(i+1,j) and b(i+1,j+1) and [b(i+2,j) or b(i+2,j+1)]
then a(i+1,j+1) ←a(i+1,j+1) or 2^2                { edge 2 }
if b(i,j) and b(i+1,j) and b(i+1,j+1)
then a(i+1,j+1) ←a(i+1,j+1) or 2^3                { edge 3 }
if b(i,j) and b(i+1,j) and [b(i,j-1) or b(i+1,j-1)]
then a(i+1,j) ←a(i+1,j) or 2^4                    { edge 4 }
if b(i,j) and b(i+1,j) and b(i,j+1)
then a(i+1,j) ←a(i+1,j) or 2^5                    { edge 5 }
if b(i,j) and b(i,j+1) and [b(i-1,j) or b(i-1,j+1)]
then a(i,j) ←a(i,j) or 2^6                        { edge 6 }
if b(i,j) and b(i,j+1) and b(i+1,j+1)
then a(i,j) ←a(i,j) or 2^7                        { edge 7 }
}
```

where:

b(i,j) is the binary value of a pixel point (i,j) and 2^k (k:0-7) is the extracted edge code.

Step3: The extracted contour edges are sorted and stored or optimized according to the application requirements. The extraction procedure is shown in Fig. 2.9.

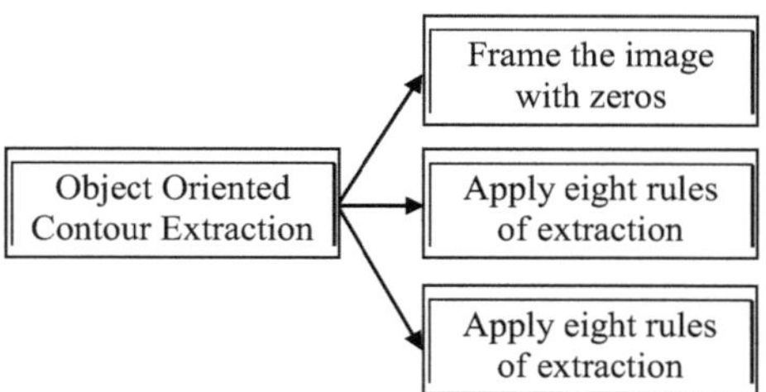

Fig. 2.9. Object-oriented Contour Extraction (OCE) method

The problem in analyzing the OCE procedure is that contours extracted from objects near the image boundary, i.e. objects within one pixel distance from the image border, are not closed and that is why the image should be framed with two background pixels to ensure the closure of the contours. Fig. (2.10a) shows that the extracted edges do not form closed contours; while Fig. (2.10b) shows that after framing the image with at least two underground pixels, all extracted contours are closed.

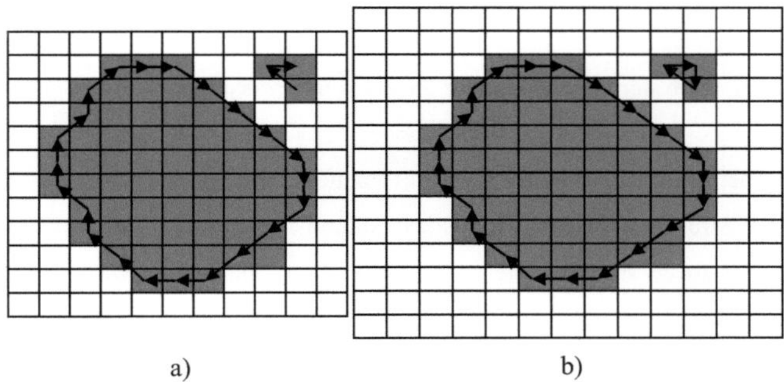

Fig. 2.10. OCE procedure (a) Without correcting the first step and (b) After correcting the first step

2.3.4 Single Step Parallel Contour Extraction 'SSPCE' Method

There are two algorithms; first one uses 8-connectivity scheme between pixels, and 8-Directional Freeman chain coding scheme is used to distinguish all eight possible line segments connecting nearest neighbors. The second algorithm uses 4-connectivity scheme between pixels, and 4-Directional Freeman chain coding scheme is used to distinguish all four possible line segments. Both algorithms uses an 3x3 pixels window structure to extract the object contours by using the central pixel to find the possible edge direction which connect the central pixel with one of the remaining pixels surrounding it. The edges can be extracted by applying the definition says that an object contour edge is a straight line connecting two neighboring pixels which have both a common neighboring object pixels which have both a common neighboring object pixel and a common neighboring underground pixel. By this definition, no edges can be extracted from the three following cases:

1- If all nine pixels are object pixel; i.e. the window is inside an object region.
2- If all nine pixels are background pixels; i.e. the window is inside a background region.
3- If the center pixel is object pixel surrounded by background pixels; i.e. most probable that the center pixel in this case is a point noise caused by image digitalization.

The following three steps are needed for the extraction procedure:

Step 1: The image is framed with zeros. This means the framing will be with first and last rows are zeros and, the first and the last column are zeros too. But why we do that?. The answer is to be sure that the edge points of the image near the borders will be extracted. In our example the binary image (D) is framed with zeros. The best way to do that we create a new matrix (E) with zeros elements has the size (n+2, m+2).
As we know from the pervious the size of the image is (22 x 22) pixels and hence the (E) matrix has size of (24 x 24) pixles. Using Matlab we can write

>> E = zeros (n+2, m+2); % matrix (E) has (24 x 24) size with zeros elements.
The next step now is to paste the binary image (D) into the E matrix. We can write
>> E (1:n, 1:m) = D (1:n, 1:m); % paste the binary image (D) into (E) matrix.
Finally we obtained the (E) binary image as shown in the Table 2.5.

Table 2.5 (framing of the binary image with zeros by adding two rows and two columns)

0	0	0	0	0	0	0	0	0	0	0	0	0	0	0	0	0	0	0	0	0	0	0	0
0	0	0	0	0	0	0	0	0	0	0	0	0	0	0	0	0	0	0	0	0	0	0	0
0	0	0	0	0	0	0	0	0	0	0	0	0	0	0	0	0	0	0	0	0	0	0	0
0	0	0	0	0	0	0	0	0	0	0	0	0	0	0	0	0	0	0	0	0	0	0	0
0	0	0	0	0	0	0	0	0	0	0	0	0	0	0	0	0	0	0	0	0	0	0	0
0	0	0	0	0	0	0	0	0	0	0	0	0	1	0	0	0	0	0	0	0	0	0	0
0	0	0	0	0	0	0	0	0	0	0	0	1	1	0	0	0	0	0	0	0	0	0	0
0	0	0	0	0	0	0	0	0	0	0	1	1	1	0	0	0	0	0	0	0	0	0	0
0	0	0	0	0	0	0	0	0	0	1	1	1	1	0	0	0	0	0	0	0	0	0	0
0	0	0	0	0	0	0	0	0	1	1	1	1	1	1	1	1	0	0	0	0	0	0	0
0	0	0	0	0	0	0	0	1	1	1	1	1	1	1	1	1	0	0	0	0	0	0	0
0	0	0	0	0	0	0	1	1	1	1	1	1	1	1	1	1	0	0	0	0	0	0	0
0	0	0	0	0	0	0	1	1	1	1	1	1	1	1	1	1	0	0	0	0	0	0	0
0	0	0	0	0	0	0	0	1	1	1	1	1	1	1	1	1	0	0	0	0	0	0	0
0	0	0	0	0	0	0	0	0	1	1	1	1	1	1	1	1	0	0	0	0	0	0	0
0	0	0	0	0	0	0	0	0	0	1	1	1	1	0	0	0	0	0	0	0	0	0	0
0	0	0	0	0	0	0	0	0	0	0	1	1	1	0	0	0	0	0	0	0	0	0	0
0	0	0	0	0	0	0	0	0	0	0	0	1	1	0	0	0	0	0	0	0	0	0	0
0	0	0	0	0	0	0	0	0	0	0	0	0	1	0	0	0	0	0	0	0	0	0	0
0	0	0	0	0	0	0	0	0	0	0	0	0	0	0	0	0	0	0	0	0	0	0	0
0	0	0		0	0	0	0	0	0	0	0	0	0	0	0	0	0	0	0	0	0	0	0
0	0	0	0	0	0	0	0	0	0	0	0	0	0	0	0	0	0	0	0	0	0	0	0
0	0	0	0	0	0	0	0	0	0	0	0	0	0	0	0	0	0	0	0	0	0	0	0
0	0	0	0	0	0	0	0	0	0	0	0	0	0	0	0	0	0	0	0	0	0	0	0

Step 2: In this step eight rules of edge extraction are applied and are coded using 8-directional freeman chain-code as shown in Listing 2.2 and Fig. 2.11.

Listing 2.2 (3x3 window)
Implementation of the eight rules for contour extraction

```
a(i,j) ←0; I = 1,2,…..,N; j = 1,2,…..,N;
for i = 2,3,…..,N-1; j = 2,3,…..,N-1;
{
if b(i,j) and b(i+1,j) and [b(i,j+1) or b(i+1,j+1)] and [ not [b(i,j-1) or
b(i+1,j-1)]]
then a(i,j) ←a(i,j) or 2^0
{ edge 0 }
if b(i,j) and b(i+1,j) and b(i+1,j-1) and [ not [b(i,j-1)]]
then a(i,j) ←a(i,j) or 2^1
{ edge 1 }
if b(i,j) and b(i,j-1) and [b(i+1,j) or b(i+1,j-1)] and [ not [b(i-1,j) or
b(i-1,j-1)]]
then a(i,j) ←a(i,j) or 2^2
{ edge 2 }
if b(i,j) and b(i,j-1) and b(i-1,j-1) and [ not [b(i-1,j)]]
then a(i,j) ←a(i,j) or 2^3
{ edge 3 }
if b(i,j) and b(i-1,j) and [b(i,j-1) or b(i-1,j-1)] and [ not [b(i,j+1) or
b(i-1,j+1)]]
then a(i,j) ←a(i,j) or 2^4
{ edge 4 }
if b(i,j) and b(i-1,j) and b(i-1,j+1) and [ not [b(i,j+1)]]
then a(i,j) ←a(i,j) or 2^5
{ edge 5 }
if b(i,j) and b(i,j+1) and [b(i-1,j) or b(i-1,j+1)] and [ not [b(i+1,j) or
b(i+1,j+1)]]
then a(i,j) ←a(i,j) or 2^6
{ edge 6 }
if b(i,j) and b(i,j+1) and b(i+1,j+1) and [ not [b(i+1,j)]]
then a(i,j) ←a(i,j) or 2^7
{ edge 7 }
```

As we can see from the Fig. 2.12 to store any edge, the current pixel (processed pixel) of the central window should be white color.
The background of the binary image is zeros and from the Fig. the code of edge 0 by the freeman chain coding is zero too. Hence the confuse will be occurred. To avoid this confusion we can code the edges as shown in the Table 2.6

Table 2.6

	Edge 0	Edge 1	Edge 2	Edge 3	Edge 4	Edge 5	Edge 6	Edge 7
Code	2^0	2^1	2^2	2^3	2^4	2^5	2^6	2^7

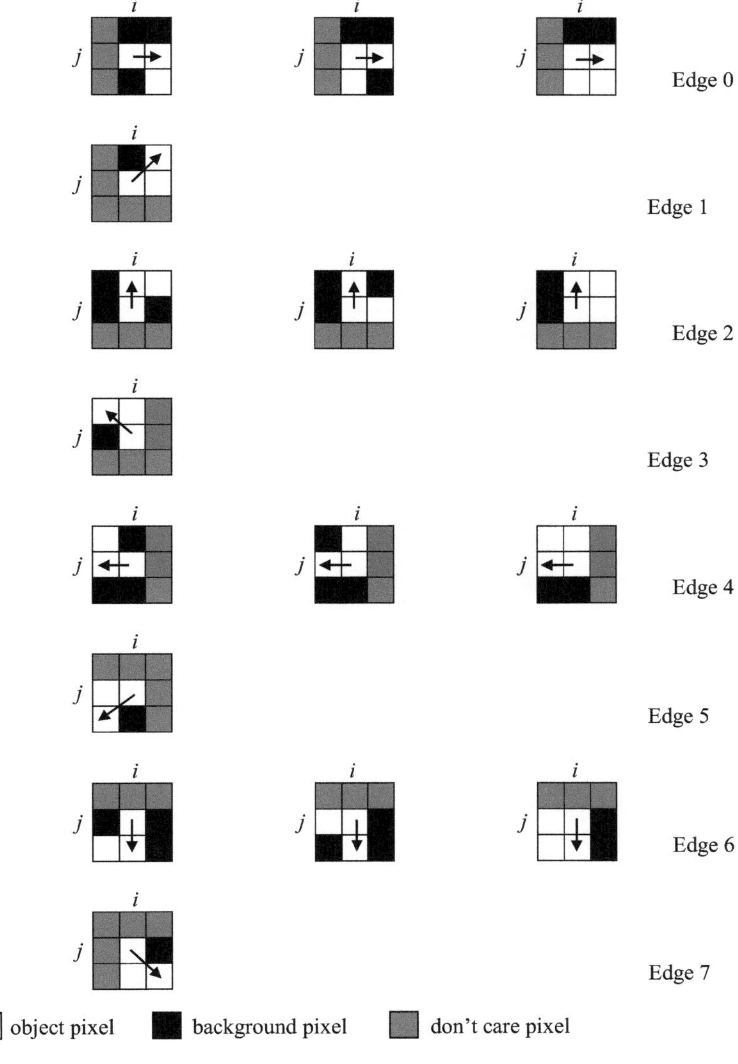

Fig. 2.11. The eight rules for closed contour extraction

Now let us apply the rules as in the Listing 2.2 and Fig.2.11 for our example (binary image D).

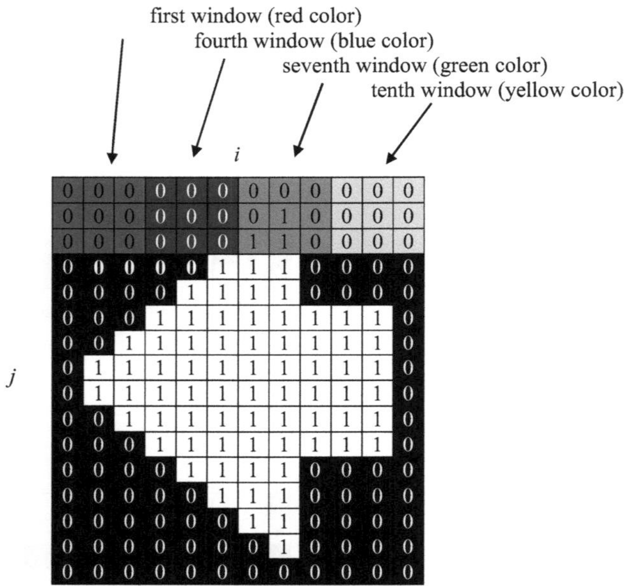

Fig. 2.12. The eight rules is applied for the binary image using SSPCE (3x3) method

The central of the first six windows as we can see from the Fig. 2.12 are zeros. Hence there is no way to detect any edge. The central of the seventh window is equals to one and the rules is applied to detect any edge may found. The edge 1 is detected by the seventh window and the code of the edge 1 ($2^6 = 64$) will take the position of the processed pixel. This means the direction of the central pixel is to the bottom ($\downarrow$).

The image scanning is done from the first row (from the left column to the right one) to the last row (from the first column to the last one).

Step 3: In this step the extracted contour edges are sorted and stored.

The codes of the final extracted edges are shown in the Fig. 2.13.

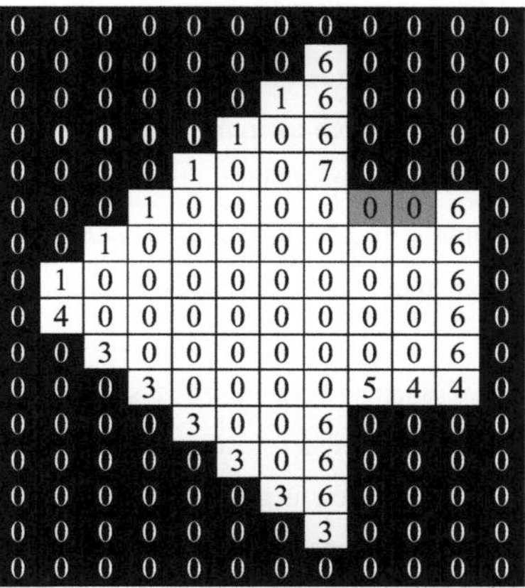

(a)

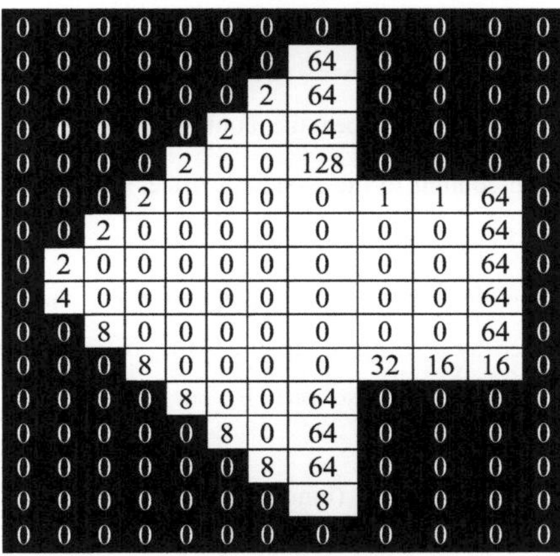

(b)

Fig. 2.13. Contour extraction using (a) Freeman chain coding, (b) Freeman modification code

4. Step4 (contour description of the extraction contour)

Generally, object representations (also called object descriptors) provide ways of describing object properties or features [17] and [18]. Finding object representations invariant to affine transformations, such as translations, rotations and linear scaling, has been a very important area in object recognition [19] and [20]. Many representations are derived from 2D shape features of 3D-objects such as object contours [17], [21], and [22]. An advantage of such an approach is that these features are relatively easy to extract and analyze.

Statistics and analysis of shapes will be an essential learning kit for statistical researches, engineers, scientists, medical researchers, and students seeking a rapid introduction to the field. It may be used as a textbook for a graduate-level special topics course in statistics and signal/image analysis, or for an intensive short course on shape analysis and modeling [23] and [24].

The description of the extracted contours normally use the chain code, i.e. the x and y co-ordinates of the starting point followed by a string of chain codes representing the contour edges. The contours are descried in either Cartesian representation, the x and y co-ordinates, or polar representation, using length of the line l from one point to the next in sequence and the angle α between every two lines [25] and [26]. There is another type of polar representation which describes the contour by using r_i (the length from a reference point inside the contour to the contour vertices) and ϕ_i (the angle between the two lines r_i and r_{i+1}).

Extracted contour is normally described using chain code, i. e. the x and y co-ordinates of the starting point followed by a string of chain cods representing the contour edges (for example Cartesian , Polar ,.....and so on). In our example we used the Cartesian representation.

To extract the contour in our example (binary image D) using Matlab programming we can write

```
>> C = SSPCE ( D );
```

The answer will be cell array [3x31]. This means that the contour have 31 points. The three roe is for the x axis, y axis and the directional code for the each point.

To show inside this contour we write

>> DD = C{1,1};% to see the first contour.

The answer will be as in the Table 2.7.

starting point is the minimum value in x-axis

Table 2.7

x	7	8	9	10	11	12	13	13	13	13	14	15	16	16	16	
y	11	10	9	8	7	6	5	6	7	8	9	9	9	10	11	
DC	1	1	1	1	1	1	6	6	6	7	0	0	6	6	6	
x	16	16	16	15	14	13	13	13	13	12	11	10	9	8	7	7
y	12	13	14	14	14	15	16	17	18	17	16	15	14	13	12	11
DC	6	6	4	4	5	6	6	6	3	3	3	3	3	3	2	1

ending point

DC - the directional code

As we can see from the table 2.7 the starting and ending points are the same to obtain the closed contour.

Now we can see the extracted contour in y-x plane (Cartesian co-ordinates). To do this we write

>> CC = DD (1:2,:); % x and y only and without directional code

To show this contour we write

>> figure; plot (CC (1,:) ,CC (2,:));% see Fig. 2.14.

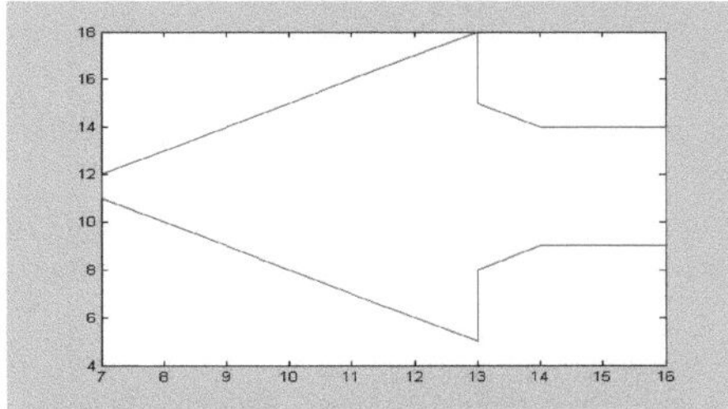

Fig. 2.14. The extracted contour from the binary image (D).

Hint: Using Matlab we can simply extract the contour from the binary image by calling the file SSPCE as

>> TA = SSPCE (BW);

where: BW is black and white image.

2.4. Contour Extraction Examples using Matlab

1. Black and White Pixels (Binary)

>> A = imread('Libya.bmp'); % reads the image in file name (Libya) into A and bmp is a string that specifies the format of the file

>> figure; imshow(A); % display the image A (see Fig.2.15)

Fig. 2.15. Binary Image (Libya map)

This is a binary image or we can say the original contour where the edges of the contour are the regions between black (zeros) and white (ones).

>> C = SSPCE(A) ; % extract the contour from the binary image (A) by calling the file SSPCE (Single Step Parallel Contour Extraction) which is used the 3-by-3 Windows.

C =

[3x798 double]

The answer is cell array with only one contour extracted. That is means that we extract one contour by three rows, x and y Cartesian-Coordinates with the direction of each of the points using 8-Directional Freeman chain coding. The contour is closed and the length of the contour is 798 points.

But we can not see the contour points from this cell. To do that the following commands is used.

>> CC = C{1,1}

CC =

44 45 45 45 46

207 206 205 204 203

1 2 2 1 1

....

....

....

Columns 791 through 795

47 46 45 45 44

214 213 212 211 210

3 3 2 3 2

Columns 796 through 798

44 44 44

209 208 207

2 2 1

>> figure; plot(CC(1,:),CC(2,:)) % display the contour CC in a private figure (Fig. 2.16)

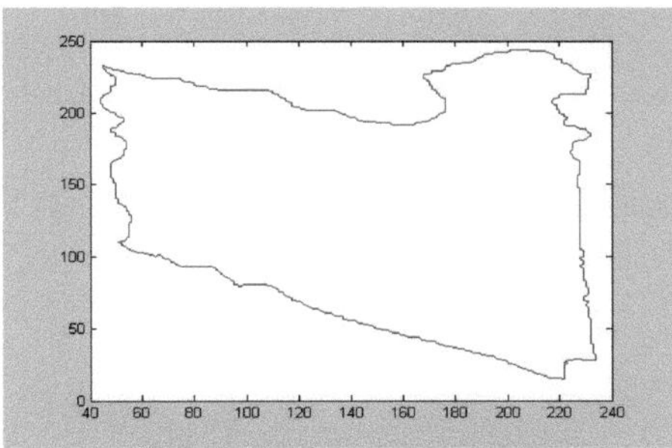

Fig. 2.16. Extracted contour plot

The comparison is made between the following three algorithms:

- SSPCE method; it will be referred to as the first algorithm (or 3x3 windows).
- OCE method; it will be referred to as the second algorithm (or 4x4 windows).

The comparison is made with respect to speed (number of arithmetic operations) for Libya map binary image.

The comparison of the Libya map contour for number of operations is shown in Table 2.8.

Table 2.8 Comparison between algorithms for the Libya map image

Method	*NO*
SSPCE	*3,389.698*
OCE	*5,779.886*

NO is the Number of Operations

The results presented in Table 2.8 show that the contour extraction using 3x3 windows gives similar contours but not identical and is faster than the second algorithm (4x4 windows) [12].

2. **Color Image**

Suppose we want to load (read) the color image (Lena) and in Matlab we can write

```
>> A = imread ('Lena.jpg'); figure; imshow (A); title ('Color Image') % see Fig. 2.17
```

Fig. 2.17. Color Image (Lena)

In Matlab we can convert the color image to the gray-level (intensity) image by using the following statement

```
>> GL = rgb2gray (A); figure; imshow (GL) % see Fig. 2.18
```

Fig. 2.18. Gray-Level Image (Lena)

In Matlab how we can obtained the binary image from the gray-level image??. To answer this question we can say we can do that by two ways.

The first way is by applied the threshold to the all image directly. If the value of the pixel is greater than the threshold value then the pixel is represented by value one (1). Otherwise the pixel is represented by zero (0). In Matlab we write

```
>> BWTH = GL > 11; figure; imshow (BWTH) % threshold value is 11 (see Fig. 2.19)
```

As we seen from the figure we did not see any image or objects. So the question is how the threshold value is chosen and the answer is by determination the minimum and maximum values of the pixels in the gray level image.

Fig. 2.19. Binary Image (Threshold is out of the Gary-level range)

```
>> min ( min ( GL ) )

ans =

   20

>> max ( max ( GL ) )

ans =

  239
```

Now for example we choose 88 as threshold value & remember that this valus should be in the range(minimum value(20) – maximum value (239)).

```
>> BWTH = GL > 88; figure; imshow ( BWTH ) % threshold value is 88 (see Fig. 2.20)
```

Fig. 2.20. Binary Image (Threshold is in the Gary-level range)

>> BWTH = GL > 122; figure; imshow (BWTH) % threshold value is 122 (see Fig. 2.21)

Fig. 2.21. Binary Image (Threshold is out of the Gary-level range)

The second way of binarization is by applied the EDGE statement which is find edges in intensity image.

First we introduce a brief description of the steps which needed to find the edges of an object in the image using Matlab.

2.5 Edge detectors

In typical images, edges characterize object boundaries and are therefore useful for segmentation, registration, and identification of objects in a scene.

Edges in images are areas with strong intensity contrasts – a jump in intensity from one pixel to the next. There are several gradient operators used for edge detection, nod variations can be obtained by rotating the kernel values divided in to two different orders [9], [13], [27], [28], [29]:

- 1st order: Roberts Cross, Prewitt, Sobel, Canny
- 2nd Order: Marr-Hildreth, zero-crossings of the second-order derivative in the gradient direction.

The goal of edge detection is to mark the points in a digital image at which the luminous intensity changes sharply. Sharp changes in image properties usually reflect important events and changes in properties of the world. These include (i) discontinuities in depth, (ii) discontinuities in surface orientation, (iii) changes in material properties and (iv) variations in scene illumination. Edge detection is a research field within image processing and computer vision, in particular within the area of feature extraction [30] and [31].

Edge detection of an image reduces significantly the amount of data and filters out information that may be regarded as less relevant, preserving the important structural properties of an image. There are many methods for edge detection, but most of them can be grouped into two categories, search-based and zero-crossing based. The search-based methods detect edges by looking for maxima and minima in the first derivative of the image, usually local directional maxima of the gradient magnitude. The zero-crossing based methods search for zero crossings in the second derivative of the image in order to find edges, usually the zero-crossings of the Laplacian or the zero-crossings of a non-linear differential expression [30] and [31].

- Edge properties

Edges may be viewpoint dependent - these are edges that may change as the viewpoint changes, and typically reflect the geometry of the scene, objects occluding one another and so on, or may be viewpoint independent - these generally reflect properties of the viewed objects such as surface markings and surface shape. In two dimensions, and higher, the concept of perspective projection has to be considered.

A typical edge might be (for instance) the border between a block of red color and a block of yellow; in contrast a line can be a small number of pixels of a different color on an otherwise unchanging background. There will be one edge on each side of the line. Edges play quite an important role in many applications of image processing. During recent years, however, substantial (and successful) research has also been made on computer vision methods that do not explicitly rely on edge detection as a pre-processing step.

- Detecting an edge

Taking an edge to be a change in intensity taking place over a number of pixels, edge detection algorithms generally compute a derivative of this intensity change. To simplify matters, we can consider the detection of an edge in one dimension. In this instance, our data can be a single line of pixel intensities. For instance, we can intuitively say that there should be an edge between the 4th and 5th pixels in the following 1-dimensional data:

5	7	6	4	152	148	149

To firmly state a specific threshold on how large the intensity change between two neighboring pixels must be for us to say that there should be an edge between these pixels is, however, not always an easy problem. Indeed, this is one of the reasons why edge detection may be a non-trivial problem unless the objects in the scene are particularly simple and the illumination conditions can be well controlled.

- Thresholding

Once we have calculated our derivative, the next stage is to apply a threshold, to determine where the result suggests an edge to be present. The lower the threshold, the more lines will be detected, and the results become increasingly susceptible to noise, and also to picking out irrelevant features from the image. Conversely a high threshold may miss subtle lines, or segmented of lines.

A commonly used compromise is thresholding with hysteresis. This method uses multiple thresholds to find edges. We begin by using the upper threshold to find the start of a line. Once we have a start point, we trace the edge's path through the image pixel by pixel, marking an edge whenever we are above the lower threshold. We stop marking our edge only when the value falls below our lower threshold. This approach makes the assumption that edges are likely to be in continuous lines, and allows us to follow a faint section of an edge we have previously seen, without meaning that every noisy pixel in the image is marked down as an edge [30] and [31].

EDGE statement takes an intensity image I as its input, and returns a binary image BW of the same size as I, with 1's where the function finds edges in I and 0's elsewhere.

BW = EDGE(I,'canny',THRESH) specifies sensitivity thresholds for the Canny method. THRESH is a two-element vector in which the first element is the low threshold, and the second element is the high threshold. If you specify a scalar for THRESH, this value is used for the high threshold and 0.4*THRESH is used for the low threshold. If you do not specify THRESH, or if THRESH is empty ([]), EDGE chooses low and high values automatically.

>> BW = edge (GL, 'Canny'); figure; imshow (BW) % see Fig. 2.22

Fig. 2.22. Edge Detector using Canny Operator (Threshold chosen automatically)

```
>> BW =  EDGE (GL, 'Canny', [ 0.01   0.2]); figure; imshow ( BW )% see Fig. 2.23
```

Thresh must be [low high], where low < high < 1.

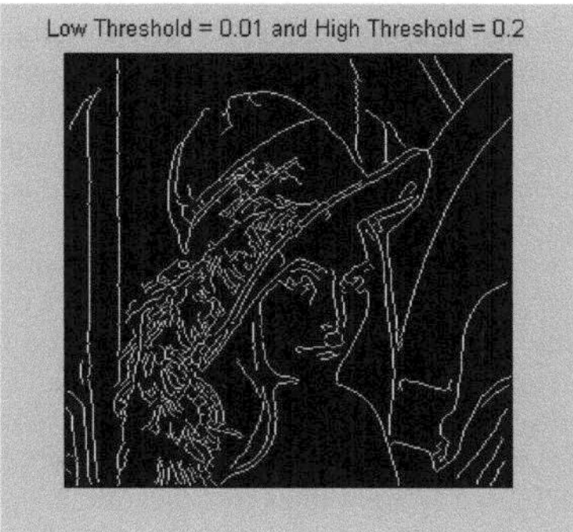

Fig. 2.23. Edge Detector using Canny Operator (Threshold chosen manually)

Instead of Canny we can use other edge detectors operators like Sobel, Prewitt, Roberts, Log and so on.

To see more details you may type

```
>> help edge
```

Chapter Three

Contour Compression Methods in Spatial Domain

3.1 Ramer method

Fits the edge points with a sequence of line segments (i.e., contour is represented as a polygon).There are several algorithms available for determining the number and location of the vertices and also to compute the polygonal approximation of a contour. The Ramer method is based on the polygonal approximation scheme. The simplest approach for the polygonal approximation is a recursive process (Splitting methods). Splitting methods work by first drawing a line from one point on the boundary to another. Then, we compute the perpendicular distance from each point along the segment to the line. If this exceeds some threshold, we break the line at the point of greatest error. We then repeat the process recursively for each of the two new lines until we don't need to break any more. For a closed contour, we can find the two points that lie farthest apart and fit two lines between them, one for one side and one for the other. Then, we can apply the recursive splitting procedure to each side. First, use a single straight line to connect the end points. Then find the edge point with the greatest distance from this straight line. Then split the straight line in two straight lines that meet at this point. Repeat this process with each of the two new lines. Recursively repeat this process until the maximum distance of any point to the poly-line falls below a certain threshold. Finally draw the lines between the vertices of an edge of the reconstructed contour to obtain the polygonal approximating contour [34].

To analyze the Ramer method let us take the following example.

Consider the closed contour (i.e the first and last points of the contour are equals) as shown in Fig 3.1. The contour is extracted using the SSPCE (Single Step Parallel Contour Extraction) method. The contour is represented by (y-x) Cartesian Co-ordinates. The sequence of the contour (*CC*) is shown in Table 3.1.

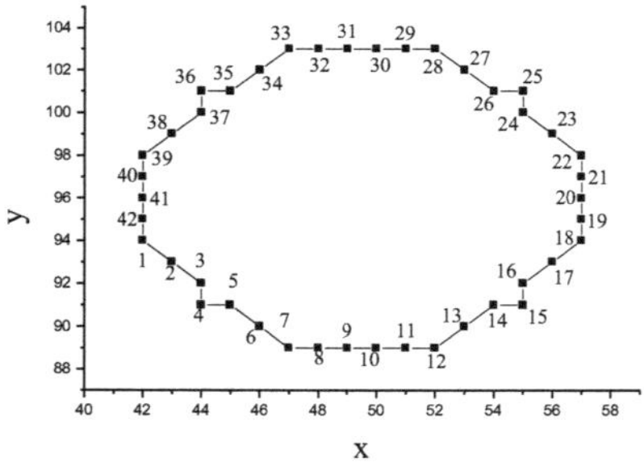

Fig. 3.1. The original contour

Table 3.1

PN	1	2	3	4	5	6	7	8	9	10	11	12	13	14
x	42	43	44	44	45	46	47	48	49	50	51	52	53	54
y	94	93	92	91	91	90	89	89	89	89	89	89	90	91
PN	15	16	17	18	19	20	21	22	23	24	25	26	27	28
x	55	55	56	57	57	57	57	57	56	55	55	54	53	52
y	91	92	93	94	95	96	97	98	99	100	101	101	102	103
PN	29	30	31	32	33	34	35	36	37	38	39	40	41	42
x	51	50	49	48	47	46	45	44	44	43	42	42	42	42
y	103	103	103	103	103	102	101	101	100	99	98	97	96	95
PN	43													
x	42													
y	94													

Where: PN is the point number (order) in the contour sequence.

The threshold value used in the example is 4.

The analysis steps are:

1. Determine the start point SP (x_1, y_1) by looking for the minimum value of x in the left side of the contour (42, 94) (see Fig. 3.2).

2. Determine the end point EP (x_2, y_2) by looking for the maximum value of x in the right side of the contour (57, 94) (see Fig. 3.2).

3. Store the end point number in the sequence of the contour as a vertex (AC = [18]).

4. Draw a straight line l_1 between the SP and EP as shown in Fig. 3.2.

As we can see from the Fig. 3.2 the closed contour is divided into two parts (lower and upper curves).

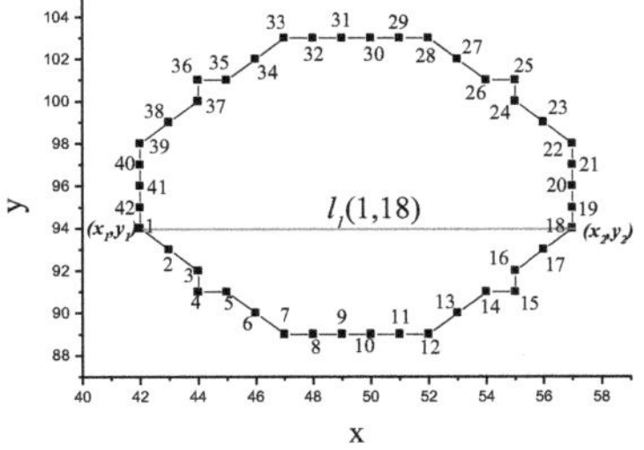

Fig. 3.2. The curve segment of straight line l_1.

Let us first consider the lower curve part. The number of operations is 1.

5. **5.** Compute the perpendicular distance from each point along the curve segment to the straight line l_1. The sequence of the points for the curve segment of the straight line l_1 is shown in Table 3.2.

Table 3.2

	SP	The distance from each point to the straight line l_1 is calculated																EP
x	42	43	44	44	45	46	47	48	49	50	51	52	53	54	55	55	56	57
y	94	93	92	91	91	90	89	89	89	89	89	89	90	91	91	92	93	94

To evaluate the distance of (x_p, y_p) point from straight line l_1 , the constants (a, b & c) for the general equation of the straight line l_1 should be determined.

$$ax_p + by_p + c = 0 \tag{3.1}$$

Where: $a = y_2 - y_1, b = x_1 - x_2 \; \& \; c = x_2y_1 - x_1y_2$

$a = y_2 - y_1 = 94 - 94 = 0,$
$b = x_1 - x_2 = 42 - 57 = -15$ and
$c = x_2y_1 - x_1y_2 = (57*94) - (42*94) = 1410$

The general equation for the straight line l_1 can be rewritten as

$$-15x_p + 1410 = 0$$

The distance of edge point (x_p, y_p) from the straight line l_1 is:

$$d = \frac{|ax_p + by_p + c|}{\sqrt{a^2 + b^2}} \tag{3.2}$$

The segment sketch of the equation is implemented in Fig. 3.3.

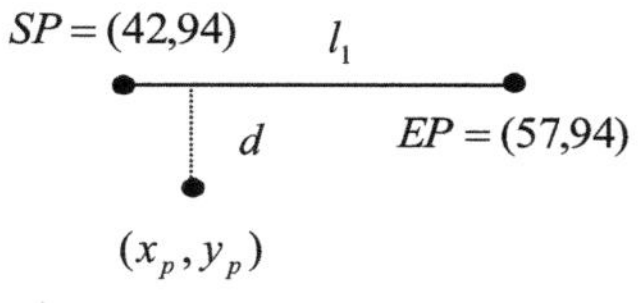

$SP = (42{,}94)$ l_1

d $EP = (57{,}94)$

(x_p, y_p)

Fig. 3.3 Computation the perpendicular distance between Point (x_p, y_p) and a line segment l_1.

Let us evaluate the perpendicular distance for each point (Table 3.2) from the straight line l_1. First the calculation is done for the first point (43, 93) (see Fig 3.4).

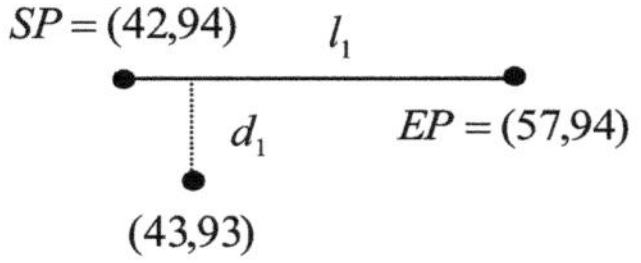

$SP = (42{,}94)$ l_1

d_1 $EP = (57{,}94)$

(43,93)

Fig. 3.4 Computation the perpendicular distance between Point (43,93) and a straight line l_1.

$$d_1 = \frac{|(0*43)+(-15*93)+1410|}{\sqrt{(0)^2+(-15)^2}} = \frac{15}{15} = 1.$$

Then calculate the distance from the second point (44, 92) to the straight line l_1.

$$d_2 = \frac{|(0*44)+(-15*92)+1410|}{\sqrt{(0)^2+(-15)^2}} = \frac{30}{15} = 2.$$

The d_2 is greater than d_1 and hence the d_2 is stored and the last distance d_1 is removed. If for example the d_1 is greater than d_2 then d_2 is removed.
The distance from the third point (45, 91) to the straight line l_1 is calculated.

$$d_3 = \frac{|(0*45)+(-15*91)+1410|}{\sqrt{(0)^2+(-15)^2}} = \frac{45}{15} = 3.$$

The d_2 is smaller than d_3 and hence the d_3 is stored and the last distance d_2 is removed. The distance from the next point in the Table 3.2 to the straight line l_1 is calculated and so on. Finally only the maximum value of perpendicular distance from this curve segment to the straight line l_1 is obtained. In our example the value of the maximum distance is 5 at the point number 7 in the sequence of the contour (i.e. the point (47, 89)) (see Fig. 3.5). The number of operations is 157.

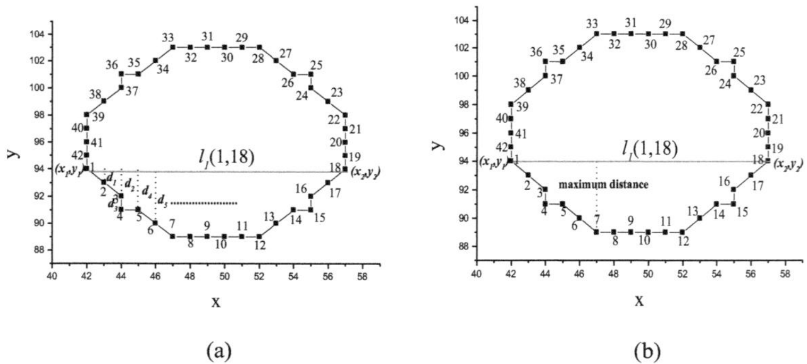

Fig. 3.5. a) Computation the perpendicular distance between points in the curve segment (Table 3.2) and a line segment l_1.
b) The maximum distance for the curve segment from the straight line l_1.

6. If this value exceeds some threshold (4), we break the line at the point of most distance from the straight line l_1 (i.e. the point (47, 89)) into two new segments (see Fig. 3.6).

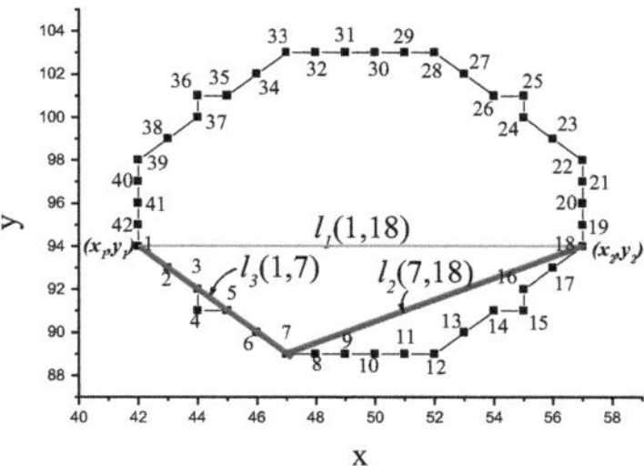

Fig. 3.6. The curve segments for the two new straight lines l_2 and l_3

7. This step is to show what happened in case of the maximum distance is exceeds the threshold value. If the perpendicular distance from this curve segment to the straight line l_1 is exceeds the threshed value (4) then there is no edge point (Table 3.2) of the curve segment for the straight line l_1 will be stored. So the starting point is shifted to the ending point number (18). This step is replaced by the step 6 in case of the maximum distance is exceeds the threshold. The vector of the approximating contour is AC = [1 18].
8. Let us back to the end of step 6. Now we have two new straight lines l_2 and l_3. First one is joint between the points numbers (7 & 18) in the sequence of the contour while the second is joint between points numbers (1 & 7) (see Fig 6).
9. Now repeat steps 5 and 6 with each of the two new straight lines l_2 and l_3.

In our example there are no points exceeds the threshold value (4) for curve segment of straight line l_2 and hence the point number 7 in the contour sequence is stored as a vertex of an edge of the approximating contour (AC = [18 7]). The number of operations after calculated the distances for the points of this curve segment from the straight line l_2 is 260.

For the straight line l_3 there are no points also exceeds the threshold and here the point number 1 in the contour sequence is stored as a vertex of an edge of the approximating contour (AC = [18 7 1]).
The number of operations after calculated the distances for the points of this curve segment from the straight line l_3 is 315. We are arrived at the end of the analysis of the lower curve part.

Now we start analysis the upper curve part and the number of operations is arrived 324.

1. Determine the start point SP (x_1, y_1) by looking for the maximum value of x in the lower right side of the contour (57, 94) (see Fig. 3.7).

2. Determine the end point EP (x_2, y_2) by looking for the minimum value of x in the lower left side of the contour (42, 94) (see Fig. 3.7).

3. Draw a straight line l_1 between the SP and EP as shown in Fig. 3.7.

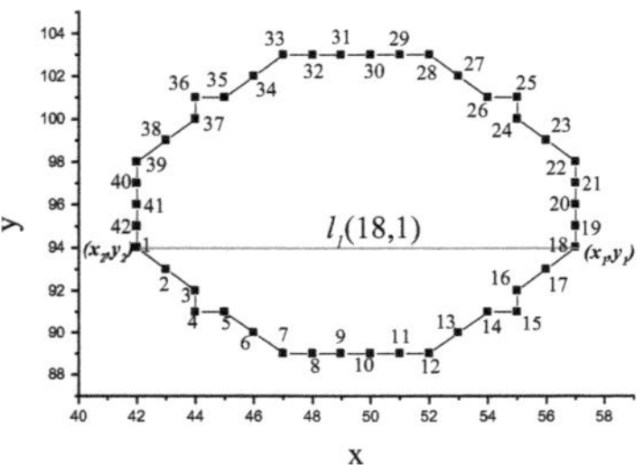

Fig. 3.7. The curve segment of straight line l_1.

4. Compute the perpendicular distance from each point along the curve segment to the straight line l_1. The sequence of the points for the curve segment of the straight line l_1 is shown in Table 3.3.

Table 3.3

	SP	The distance from each point to the straight line l_1 is calculated									EP
x	57	57	57	57	57	56	55	55	54	53	-
y	94	95	96	97	98	99	100	101	101	102	-
x	-	52	51	50	49	48	47	46	45	44	-
y	-	103	103	103	103	103	103	102	101	101	-
x	-	44	43	42	42	42	42	-	-	-	92
y	-	100	99	98	97	96	95	-	-	-	94

To evaluate the distance of (x_p, y_p) point from straight line l_1, the constants (a, b & c) for the general equation of the straight line l_1 should be determined.

$a = y_2 - y_1 = 94 - 94 = 0,$

$b = x_1 - x_2 = 57 - 42 = 15$ and
$c = x_2 y_1 - x_1 y_2 = (42*94) - (57*94) = -1410$

The general equation for the straight line l_1 can be rewritten as

$$15x_p - 1410 = 0$$

Let us evaluate the perpendicular distance for each point (Table 3.3) from the straight line l_1. First the calculation is done for the first point (57, 95) (see Fig 3.8). The d_1 is the distance between the point number 19 and the straight line (see Fig. 3.8).

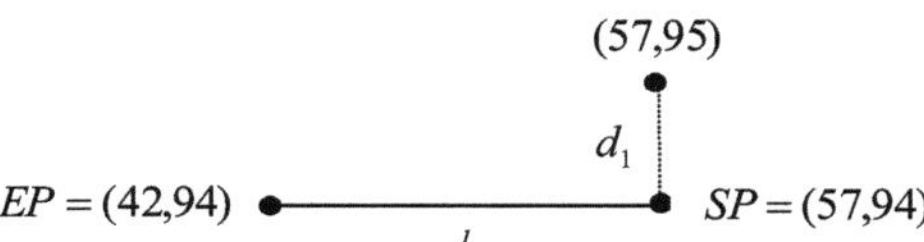

Fig. 3.8. Computation the perpendicular distance between Point (57,95) and a straight line l_1.

$$d_1 = \frac{|(0*57) + (15*95) - 1410|}{\sqrt{(0)^2 + (15)^2}} = \frac{15}{15} = 1.$$

Then calculate the distance from the second point (57, 96) to the straight line l_1. The d_2 is the distance between the point number 20 and the straight line l_1.

$$d_2 = \frac{|(0*57) + (15*96) - 1410|}{\sqrt{(0)^2 + (15)^2}} = \frac{30}{15} = 2.$$

The d_2 is greater than d_1 and hence the d_2 is stored and the last distance d_1 is removed. If for example the d_1 is greater than d_2 then d_2 is removed.
The distance from the third point (57, 97) to the straight line l_1 is calculated.

$$d_3 = \frac{|(0*57)+(15*97)-1410|}{\sqrt{(0)^2+(15)^2}} = \frac{45}{15} = 3.$$

The d_2 is smaller than d_3 and hence the d_3 is stored and the last distance d_2 is removed. The distance from the next point in the Table 3.3 to the straight line l_1 is calculated and so on. Finally only the maximum value of perpendicular distance from this curve segment to the straight line l_1 is obtained. In our example the value of the maximum distance is 9 at the point number 28 in the sequence of the contour (i.e. the point (52, 103)) (see Fig. 3.9).

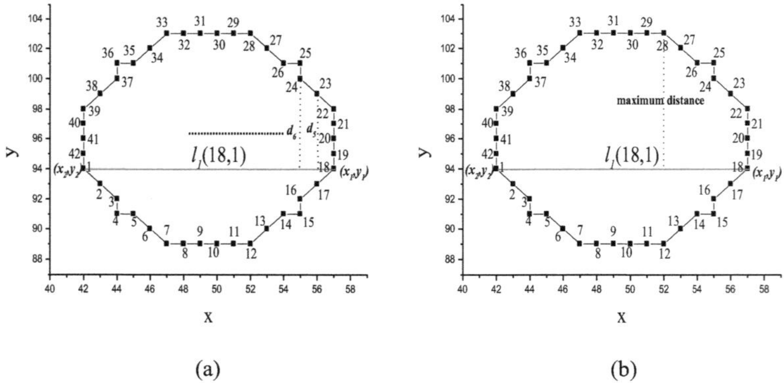

Fig. 3.9. a) Computation the perpendicular distance between points in the curve segment (Table 3.2) and a line segment l_1.
b) The maximum distance for the curve segment from the straight line l_1.

The number of operations after calculated the distances for the points from the straight line l_1 is 567.

5. If the maximum distance (9) value exceeds the threshold value (4), we break the line at the point of most distance from the straight line l_1 (i.e. the point (52, 103)) into two new segments (see Fig 3.10).

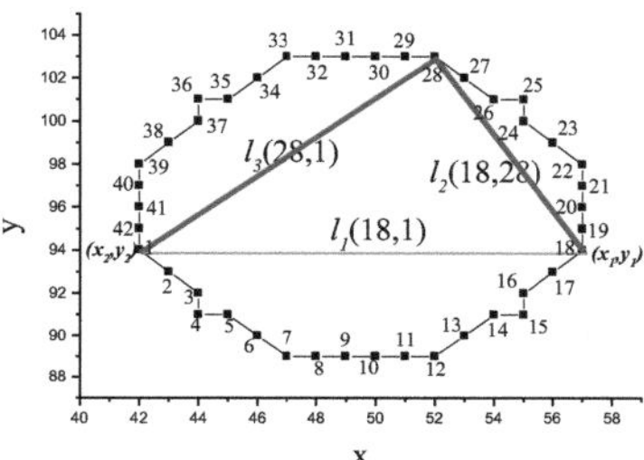

Fig. 3.10. The curve segments for the two new straight lines l_2 and l_3.

6. Now repeat steps 4 and 5 with each of the two new straight lines l_2 and l_3.

In our example there are no points exceeds the threshold (4 is our choice) for curve segment of straight line l_2 and hence the point number 28 in the contour sequence is stored as a vertex of an edge of the approximating contour (AC = [18 7 1 28]). The number of operations after calculated the distances for the points of this curve segment from the straight line l_2 is 707.

For the straight line l_3 there are no points also exceeds the threshold and here the point number 43 in the contour sequence is stored as a vertex of an edge of the approximating contour (AC = [18 7 1 28 43]). The number of operations after calculated the distances for the points of this curve segment from the straight line l_3 is 801.

The final vertices in AC are sorted to get finally the following contour points:
AC = [1 7 18 28 43]
The original contour and the approximating contour (AC) are shown in Fig. 3.11.

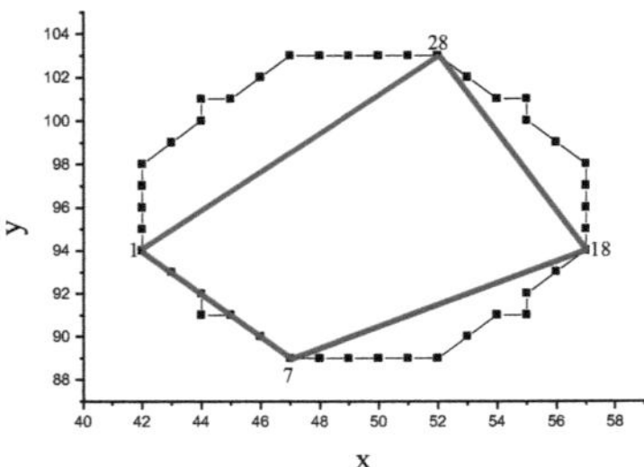

Fig. 3.11. The Original and approximated contours

The number of operations is 803.

As we can see the original contour contains 43 points while the approximated contour contains only 5 points. This ratio between these values is $(\frac{5}{43}) = 0.1163$ which is means that the 0.1163 of the contour information is reconstructed the original contour while the rest is rejected.

- Contours Compression Measures

To evaluate its compression ability, the following compression ratio was introduced

$$CR = \frac{(L_{CC} - L_{AC})}{L_{CC}} \cdot 100\% \tag{3.3}$$

Where:

L_{CC} - length of the input contour (43 in our example);

L_{AC} - length of the approximating polygon (5 in our example).

The mean square error (*MSE*) and signal-to-noise ratio (*SNR*) criterions were used to evaluate the distortion introduced during the approximating procedure. The *MSE* criterion is defined by the following equation:

$$MSE = \frac{1}{l_{CC}} \sum_{i=1}^{l_{CC}} d_i^2 \tag{3.4}$$

Where:

d_i - distance between i vertex and the input contour line.

The *SNR* is defined by the following formula:

$$SNR = -10 * \log_{10}\left(\frac{MSE}{VAR}\right) \tag{3.5}$$

Let us back to our last example. So we have the original and approximated contours. The compression ratio for our example is $CR = \frac{(43-5)}{43} * 100\% = 88.37\%$

The mean square error (*MSE*) and signal-to-noise ratio (*SNR*) criterions were used to evaluate the distortion introduced during the approximation procedure.

In our example the approximated contour has four straight lines:
The straight line l_1 between the points number 1 and 7, the straight line l_2 between the points 7 and 18, the straight line l_3 between the point's number 18 and 28 and the straight line l_4 between the points number 28 and 43.
Let us consider the first straight line l_1 and calculate the distance from each edge points in the original contour to the straight line l_1 using the equations number 1 and 2 (see Fig 3.12)

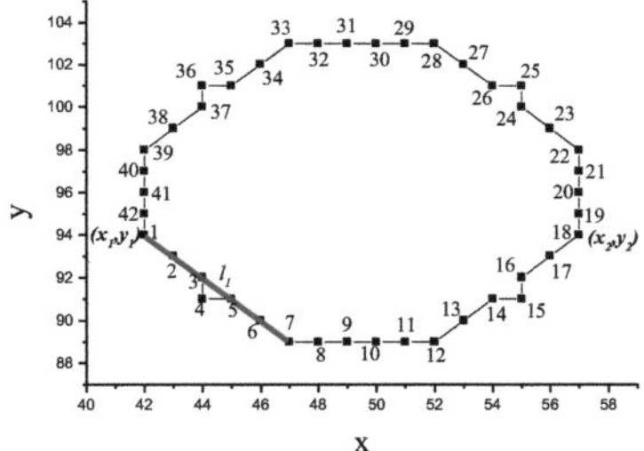

Fig. 3.12. Computation the distances from each of the points (1 to 7) to the straight line l_1.

To do that the constants for the straight line l_1 is calculated.
The start point for this line is $(x_1, y_1) = (42,94)$ and the end point is $(x_2, y_2) = (47,89)$.

$a = y_2 - y_1 = 89 - 94 = -5$,
$b = x_1 - x_2 = 42 - 47 = -5$ and

$c = x_2 y_1 - x_1 y_2 = (47*94) - (42*89) = 680$

The general equation for the straight line l_1 is $-5x_p - 5y_p + 680 = 0$

It is clear from the Fig. 3.12 the distances from each point to the straight line l_1 is 0 except the point number 4. To evaluate the distance for the point number 4 (44, 91) the following equation has been used

$$d_4 = \frac{|(-5*44) - (5*91) + 680|}{\sqrt{(-5)^2 + (-5)^2}} = \frac{5}{7.0711} = 0.7071$$

The squared of this distance is 0.5.
Therefore we obtained the all squared distances for the edge points of this curve segment of the straight line l_1 which is (0.5).
Repeat the calculations to find the distances from each edge points of each curve segment in the original contour to the each straight line l_2, l_3 and l_4 (see Fig 3.13).
The all squared distances obtained by the straight line l_2 is 20.2.
The all squared distances obtained by the straight line l_3 is 14.3302.
The all distances obtained by the straight line l_4 is 96.0718.

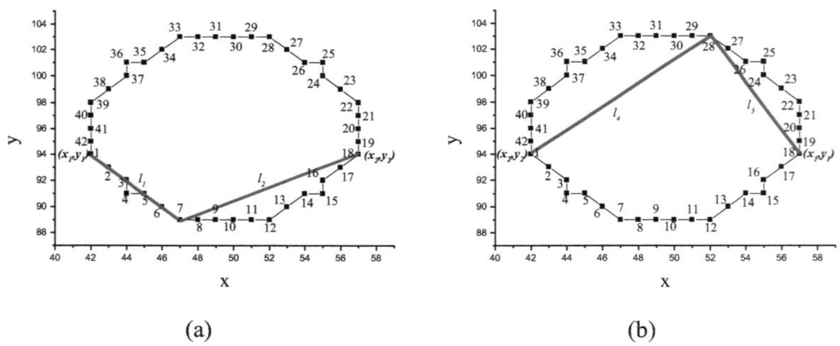

Fig. 3.13. The line segments obtained by a) Lower curve part b) Upper curve part

The *MSE* for our example is the summation of the squared distances for the all points from the polygonal approximating straight lines divided by the original contour length

$$MSE = \frac{1}{43}(0.5 + 20.2 + 14.3302 + 96.0718) = 3.0489$$

The variance, often denoted by σ^2. The variance is a measure of how spread out a distribution is. It is computed as the average squared deviation of each number from its mean. Numerically, the variance equals the average of the several squared deviations from the mean.

In our example the mean value (μ) of the original contour is 72.6395.

The general formula of the variance is

$$\sigma^2 = \frac{\sum (X - \mu)^2}{N} \tag{3.6}$$

Where μ is the mean and N is the number of scores.

In our example the N is equals to (number of rows * number of columns) $= (2*43) =$ 86.

The variance of the input of the original contour is

$$\sigma^2 = \frac{\sum_{n=1}^{2} \sum_{m=1}^{43} (CC_{nm} - 72.6395)^2}{86} = 569.6956$$

Where CC_{nm} is the input of the original contour.

Finally the *SNR* for our example is

$$SNR = -10 * \log_{10}\left(\frac{3.0489}{569.6956}\right) = 22.7150$$

Hint: Using Matlab we can simply make compression for the extracted contour by calling the file AproxRamer as

```
>> TA = AproxRamer ( CC, 4 );
```

where: CC is sequence of the input for the contour and 4 is the threshold value.

3.2 Tangent method

There are several algorithms available for determining the number and location of the vertices and also to compute the polygonal approximation of a contour. The Tangent method is based on the polygonal approximation scheme. The simplest approach for the polygonal approximation is a recursive process (Splitting methods). Splitting methods work by first drawing a line from one point on the boundary to another. Then, we compute the perpendicular distance from each point along the segment to the line. The fit criterion of the TM algorithm is the tangent of an angle between two straight lines, called opening and closing lines. These two lines determine the width of analysed segment - a set of input sequence points which is going to be replaced by an edge of the approximating polygon [35] and [36].

The fit criterion of the TM algorithm is illustrated in Fig. 3.14.

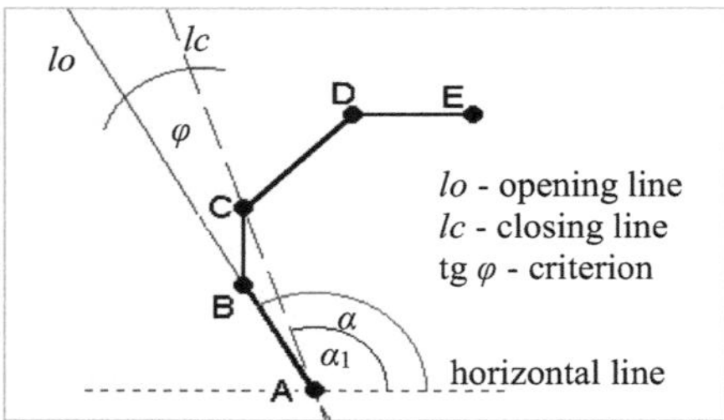

Fig. 3.14. Fit criterion of the TM algorithm

The opening line is passing through the first two points of a new segment under the investigation. The first point of each segment is the starting point SP. The SP point and the third point - the ending point EP - of the investigated segment determine the closing line. If the tangent of the φ angle is less than a given threshold p then EP point is shifted to the next point and the new closing line is drawn.

To analyze the Tangent method let us take the following example.

Consider the closed contour (i.e. the first and last points of the contour are equals) as shown in Fig 3.15. The contour is extracted using the SSPCE (Single Step Parallel Contour Extraction) method. The contour is represented by (y-x) Cartesian Co-ordinates. The sequence of the contour (*CC*) is shown in Table 3.4.

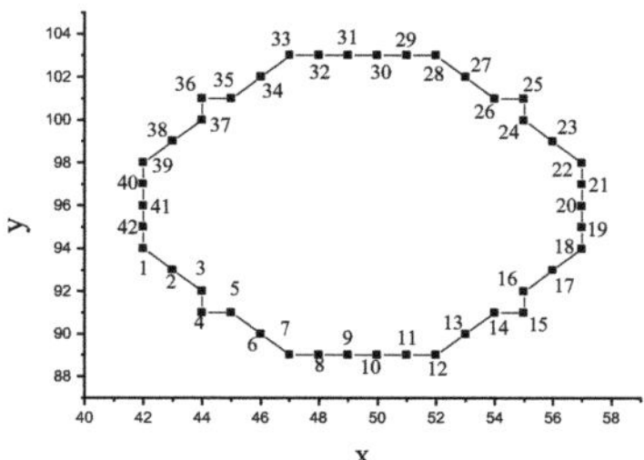

Fig. 3.15. The original contour

Table 3.4

PN	1	2	3	4	5	6	7	8	9	10	11	12	13	14
x	42	43	44	44	45	46	47	48	49	50	51	52	53	54
y	94	93	92	91	91	90	89	89	89	89	89	89	90	91
PN	15	16	17	18	19	20	21	22	23	24	25	26	27	28
x	55	55	56	57	57	57	57	57	56	55	55	54	53	52
y	91	92	93	94	95	96	97	98	99	100	101	101	102	103
PN	29	30	31	32	33	34	35	36	37	38	39	40	41	42
x	51	50	49	48	47	46	45	44	44	43	42	42	42	42
y	103	103	103	103	103	102	101	101	100	99	98	97	96	95
PN	43													
x	42													
y	94													

where: PN is the point number (order) in the contour sequence.

The threshold value used in the example is 4.

The analysis steps are:

1. Determine the start point SP (x_1, y_1) by looking for the minimum value of x in the left side of the contour (42, 94) (see Fig. 3.16).

2. Determine the end point second point (B) (x_2, y_2) by looking for the maximum value of x in the right side of the contour (43, 93) (see Fig. 3.16).

3. Determine the end point EP (x_3, y_3) by looking for the maximum value of x in the right side of the contour (44, 92) (see Fig. 3.16).

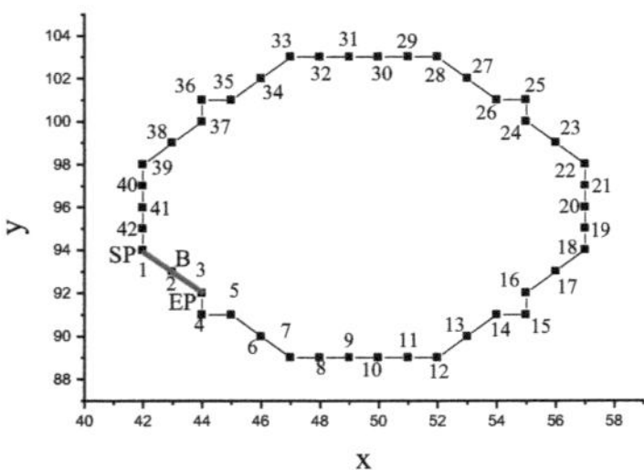

Fig. 3.16. The first segment

4. Compute the slope of the closing line (SP and EP are the end points).

```
>> m1 = (EP(2) - SP(2)) / (EP(1) - SP(1));
m1=
    -1
```

5. Compute the slope of the opening line (SP and B are the end points).

```
>> m2 = (B(2) - SP(2)) / (B(1) - SP(1));
m2=
    -1
```

6. Compute the tangent of the angle between the opening and closing lines. Assume that the angle is

$$\tan(\varphi) = \frac{m_1 - m_2}{1 + (m_1 * m_2)} \tag{3.7}$$

>> $\tan(\varphi) = \dfrac{m_1 - m_2}{1 + (m_1 * m_2)}$;

$\tan(\varphi) =$

0

7. If the value of $\tan(\varphi)$ exceeds some threshold (let us say 40 degree which equals to tan ((40*pi)/180) = 0.8391 radians). The $\tan(\varphi)$ is smaller than the threshold and then EP point is shifted to the next point in the (44, 91) sequence of the contour (see Fig. 3.17). The SP point is (42, 94), B point is (43, 93) and EP point is (44, 91).

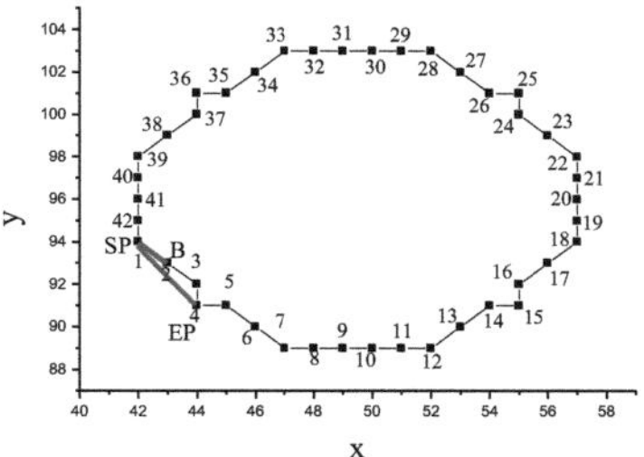

Fig. 3.17. The second segment

8. m1 is -1.5 and m2 is the same (-1) and $\tan(\varphi)$ is equals to -0.2.
9. If the value of $\tan(\varphi)$ exceeds some threshold (0.8391). The $\tan(\varphi)$ is smaller than the threshold and then EP point is shifted to the next point in the (44, 91) sequence of the contour (see Fig. 3.17). The SP point is (42, 94), B point is (43, 93) and EP point is (44, 91).
10. Compare the $\tan(\varphi)$

```
>> m = tan (B(2) - SP(2)) / (B(1) - SP(1));
m2=
    -1
```

11. Draw a straight line l_1 between the SP and EP as shown in Fig. 3.18.

As we can see from the Fig. 3.18 the closed contour is divided into two parts (lower and upper curves).

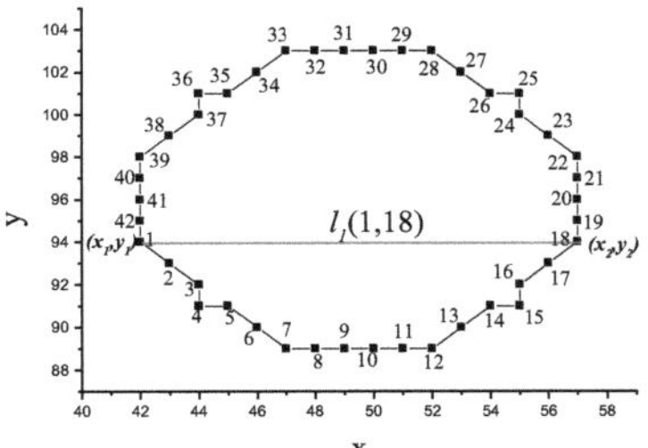

Fig. 3.18. The curve segment of straight line l_1.

Let us first consider the lower curve part. The number of operations is 1.

12. Compute the perpendicular distance from each point along the curve segment to the straight line l_1. The sequence of the points for the curve segment of the straight line l_1 is shown in Table 3.5.

Table 3.5

	SP	The distance from each point to the straight line l_1 is calculated															EP
x	42	43	44	44	45	46	47	48	49	50	51	52	53	54	55	55	56	57
y	94	93	92	91	91	90	89	89	89	89	89	89	90	91	91	92	93	94

To evaluate the distance of (x_p, y_p) point from straight line l_1 , the constants (a, b & c) for the general equation of the straight line l_1 should be determined using equation (3.1).

$a = y_2 - y_1 = 94 - 94 = 0,$
$b = x_1 - x_2 = 42 - 57 = -15$ and
$c = x_2 y_1 - x_1 y_2 = (57*94) - (42*94) = 1410$

The general equation for the straight line l_1 can be rewritten as

$$-15x_p + 1410 = 0$$

The distance of edge point (x_p, y_p) from the straight line l_1 is calculated using equation (3.2) as shown in Fig. 3.19.

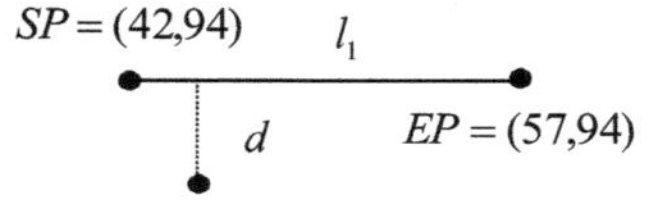

$SP = (42{,}94)$ l_1 d $EP = (57{,}94)$

Fig. 3.19. Compı (x_p, y_p) ›erpendicular distance between Point (x_p, y_p) and a line segment l_1.

Let us evaluate the perpendicular distance for each point (Table 3.5) from the straight line l_1. First the calculation is done for the first point (43, 93) (see Fig.3.20).

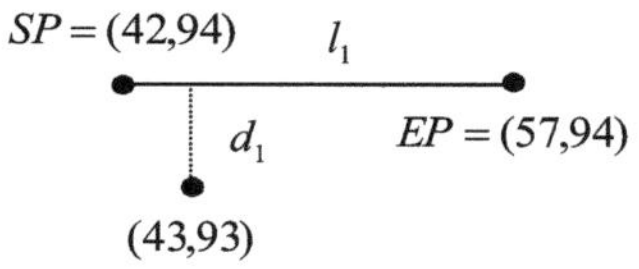

$SP = (42,94)$ l_1 d_1 $EP = (57,94)$ (43,93)

Fig. 3.20. Computation the perpendicular distance between Point (43,93) and a straight line l_1.

$$d_1 = \frac{|(0*43)+(-15*93)+1410|}{\sqrt{(0)^2+(-15)^2}} = \frac{15}{15} = 1.$$

Then calculate the distance from the second point (44, 92) to the straight line l_1.

$$d_2 = \frac{|(0*44)+(-15*92)+1410|}{\sqrt{(0)^2+(-15)^2}} = \frac{30}{15} = 2.$$

The d_2 is greater than d_1 and hence the d_2 is stored and the last distance d_1 is removed. If for example the d_1 is greater than d_2 then d_2 is removed.

The distance from the third point (45, 91) to the straight line l_1 is calculated.

$$d_3 = \frac{|(0*45)+(-15*91)+1410|}{\sqrt{(0)^2+(-15)^2}} = \frac{45}{15} = 3.$$

The d_2 is smaller than d_3 and hence the d_3 is stored and the last distance d_2 is removed. The distance from the next point in the Table 3.5 to the straight line l_1 is calculated and so on. Finally only the maximum value of perpendicular distance from this curve segment to the straight line l_1 is obtained. In our example the value of the maximum distance is 5 at the point number 7 in the sequence of the contour (i.e. the point (47, 89)) (see Fig. 3.21). The number of operations is 157.

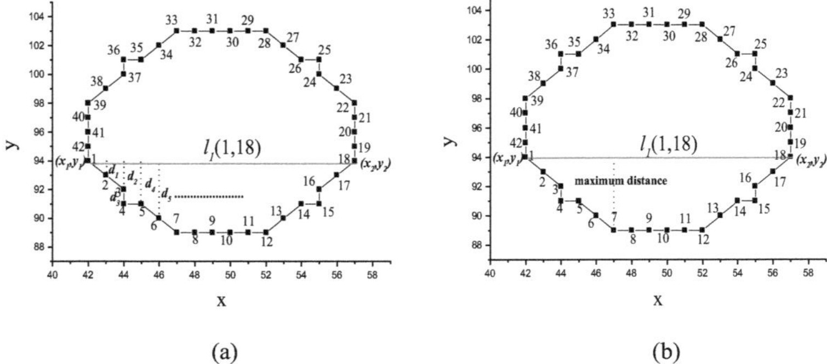

(a) (b)

Fig. 3.21. a) Computation the perpendicular distance between points in the curve segment (Table 3.5) and a line segment l_1.
b) The maximum distance for the curve segment from the straight line l_1.

13. If this value exceeds some threshold (4), we break the line at the point of most distance from the straight line l_1 (i.e. the point (47, 89)) into two new segments (see Fig. 3.22).

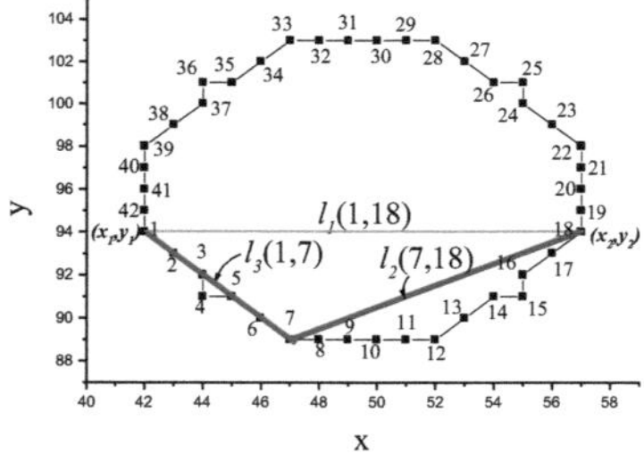

Fig. 3.22. The curve segments for the two new straight lines l_2 and l_3.

14. This step is to show what happened in case of the maximum distance is exceeds the threshold value. If the perpendicular distance from this curve segment to the straight line l_1 is exceeds the threshed value (4) then there is no edge point (Table 3.5) of the curve segment for the straight line l_1 will be stored. So the starting point is shifted to the ending point number (18). This step is replaced by the step 6 in case of the maximum distance is exceeds the threshold. The vector of the approximating contour is AC = [1 18].

15. Let us back to the end of step 6. Now we have two new straight lines l_2 and l_3. First one is joint between the points numbers (7 & 18) in the sequence of the contour while the second is joint between points numbers (1 & 7) (see Fig 3.22.

16. Now repeat steps 5 and 6 with each of the two new straight lines l_2 and l_3.

In our example there are no points exceeds the threshold value (4) for curve segment of straight line l_2 and hence the point number 7 in the contour sequence is stored as a vertex of an edge of the approximating contour (AC = [18 7]). The number of operations after calculated the distances for the points of this curve segment from the straight line l_2 is 260.

For the straight line l_3 there are no points also exceeds the threshold and here the point number 1 in the contour sequence is stored as a vertex of an edge of the approximating contour (AC = [18 7 1]).
The number of operations after calculated the distances for the points of this curve segment from the straight line l_3 is 315. We are arrived at the end of the analysis of the lower curve part.

Now we start analysis the upper curve part and the number of operations is arrived 324.

7. Determine the start point SP (x_1, y_1) by looking for the maximum value of x in the lower right side of the contour (57, 94) (see Fig. 3.23).

8. Determine the end point EP (x_2, y_2) by looking for the minimum value of x in the lower left side of the contour (42, 94) (see Fig. 3.23).

9. Draw a straight line l_1 between the SP and EP as shown in Fig. 3.23.

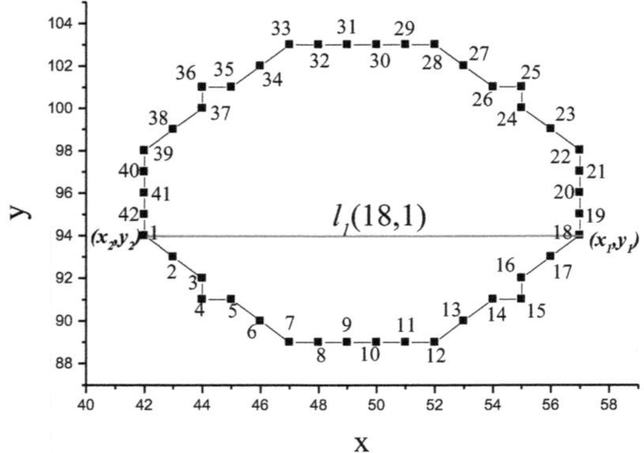

Fig. 3.23. The curve segment of straight line l_1.

10. Compute the perpendicular distance from each point along the curve segment to the straight line l_1. The sequence of the points for the curve segment of the straight line l_1 is shown in Table 3.6.

Table 3.6

	SP	The distance from each point to the straight line l_1 is calculated									EP
x	57	57	57	57	57	56	55	55	54	53	-
y	94	95	96	97	98	99	100	101	101	102	-
x	-	52	51	50	49	48	47	46	45	44	-
y	-	103	103	103	103	103	103	102	101	101	-
x	-	44	43	42	42	42	42	-	-	-	92
y	-	100	99	98	97	96	95	-	-	-	94

To evaluate the distance of (x_p, y_p) point from straight line l_1, the constants (a, b & c) for the general equation of the straight line l_1 should be determined.

$a = y_2 - y_1 = 94 - 94 = 0,$

$b = x_1 - x_2 = 57 - 42 = 15$ and

$c = x_2 y_1 - x_1 y_2 = (42 * 94) - (57 * 94) = -1410$

The general equation for the straight line l_1 can be rewritten as

$$15x_p - 1410 = 0$$

Let us evaluate the perpendicular distance for each point (Table 3.6) from the straight line l_1. First the calculation is done for the first point (57, 95) (see Fig 3.24). The d_1 is the distance between the point number 19 and the straight line l_1.

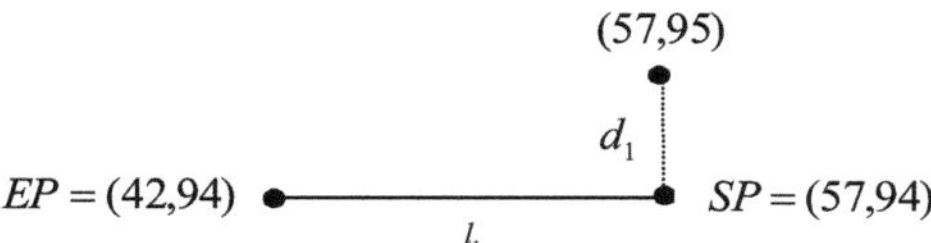

Fig. 3.24. Computation the perpendicular distance between Point (57,95) and a straight line l_1.

$$d_1 = \frac{|(0*57) + (15*95) - 1410|}{\sqrt{(0)^2 + (15)^2}} = \frac{15}{15} = 1.$$

Then calculate the distance from the second point (57, 96) to the straight line l_1. The d_2 is the distance between the point number 20 and the straight line l_1.

$$d_2 = \frac{|(0*57) + (15*96) - 1410|}{\sqrt{(0)^2 + (15)^2}} = \frac{30}{15} = 2.$$

The d_2 is greater than d_1 and hence the d_2 is stored and the last distance d_1 is removed. If for example the d_1 is greater than d_2 then d_2 is removed.
The distance from the third point (57, 97) to the straight line l_1 is calculated.

$$d_3 = \frac{|(0*57) + (15*97) - 1410|}{\sqrt{(0)^2 + (15)^2}} = \frac{45}{15} = 3.$$

The d_2 is smaller than d_3 and hence the d_3 is stored and the last distance d_2 is removed. The distance from the next point in the Table 3.6 to the straight line l_1 is calculated and so on. Finally only the maximum value of perpendicular distance from this curve segment to the straight line l_1 is obtained. In our example the value of the maximum distance is 9 at the point number 28 in the sequence of the contour (i.e. the point (52, 103)) (see Fig. 3.25).

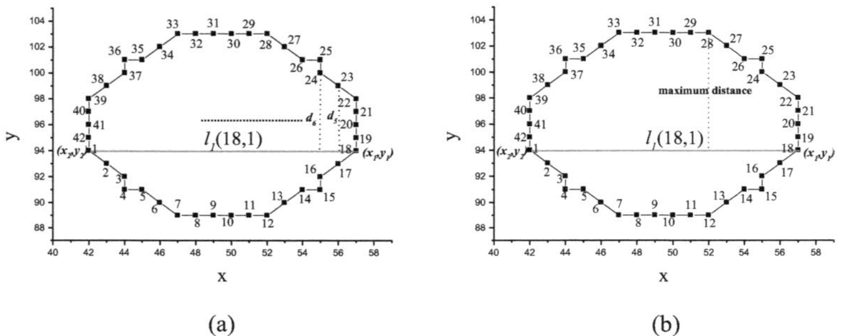

Fig. 3.25. a) Computation the perpendicular distance between points in the curve segment (Table 3.6) and a line segment l_1.
b) The maximum distance for the curve segment from the straight line l_1.

The number of operations after calculated the distances for the points from the straight line l_1 is 567.

11. If the maximum distance (9) value exceeds the threshold value (4), we break the line at the point of most distance from the straight line l_1 (i.e. the point (52, 103)) into two new segments (see Fig. 3.26).

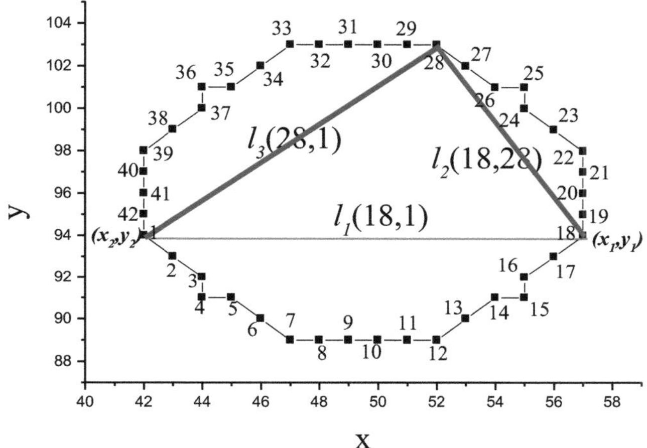

Fig. 3.26. The curve segments for the two new straight lines l_2 and l_3.

12. Now repeat steps 4 and 5 with each of the two new straight lines l_2 and l_3.

In our example there are no points exceeds the threshold (4 is our choice) for curve segment of straight line l_2 and hence the point number 28 in the contour sequence is stored as a vertex of an edge of the approximating contour (AC = [18 7 1 28]). The number of operations after calculated the distances for the points of this curve segment from the straight line l_2 is 707.

For the straight line l_3 there are no points also exceeds the threshold and here the point number 43 in the contour sequence is stored as a vertex of an edge of the approximating contour (AC = [18 7 1 28 43]). The number of operations after calculated the distances for the points of this curve segment from the straight line l_3 is 801.

The final vertices in AC are sorted to get finally the following contour points:
AC = [1 7 18 28 43]
The original contour and the approximating contour (AC) are shown in Fig. 3.27.

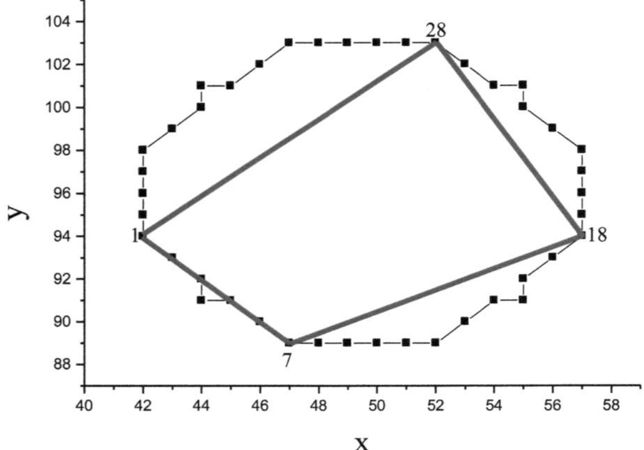

Fig. 3.27. The Original and approximated contours.

The number of operations is 803.

As we can see the original contour contains 43 points while the approximated contour contains only 5 points. This ratio between these values is $(\frac{5}{43}) = 0.1163$ which is means that the 0.1163 of the contour information is reconstructed the original contour while the rest is rejected.

The compression ratio using equation (3.3) for our example is $CR = \frac{(43-5)}{43} * 100\% = 88.37\%$

In our example the approximated contour has four straight lines:
The straight line l_1 between the points number 1 and 7, the straight line l_2 between the points 7 and 18, the straight line l_3 between the point's number 18 and 28 and the straight line l_4 between the points number 28 and 43.
Let us consider the first straight line l_1 and calculate the distance from each edge points in the original contour to the straight line l_1 using the equations number 1 and 2 (see Fig 3.28)

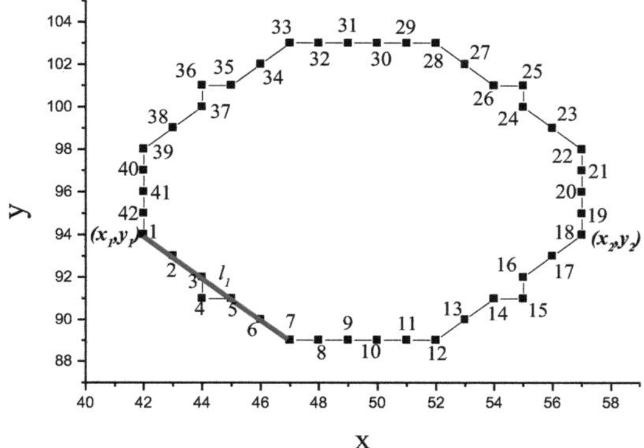

Fig. 3.28. Computation the distances from each of the points (1 to 7) to the straight line l_1.

To do that the constants for the straight line l_1 is calculated.
The start point for this line is $(x_1, y_1) = (42,94)$ and the end point is $(x_2, y_2) = (47,89)$.

$a = y_2 - y_1 = 89 - 94 = -5,$
$b = x_1 - x_2 = 42 - 47 = -5$ and
$c = x_2 y_1 - x_1 y_2 = (47 * 94) - (42 * 89) = 680$
The general equation for the straight line l_1 is $-5x_p - 5y_p + 680 = 0$.

It is clear from the Fig. 3.28 the distances from each point to the straight line l_1 is 0 except the point number 4. To evaluate the distance for the point number 4 (44, 91) the following equation has been used

$$d_4 = \frac{|(-5 * 44) - (5 * 91) + 680|}{\sqrt{(-5)^2 + (-5)^2}} = \frac{5}{7.0711} = 0.7071$$

The squared of this distance is 0.5.
Therefore we obtained the all squared distances for the edge points of this curve segment of the straight line l_1 which is (0.5).

Repeat the calculations to find the distances from each edge points of each curve segment in the original contour to the each straight line l_2, l_3 and l_4 (see Fig 3.29).
The all squared distances obtained by the straight line l_2 is 20.2.
The all squared distances obtained by the straight line l_3 is 14.3302.
The all distances obtained by the straight line l_4 is 96.0718.

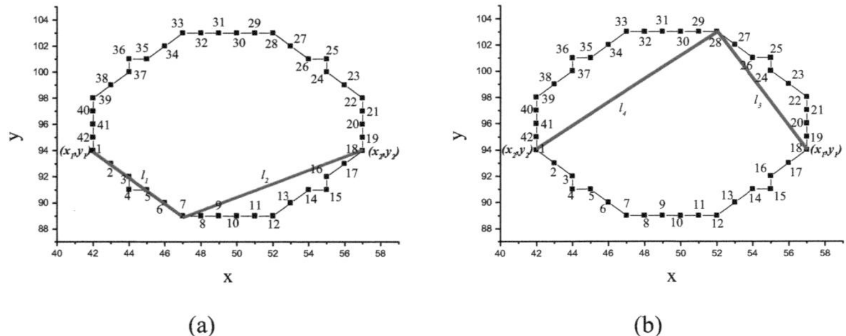

Fig. 3.29. The line segments obtained by a) Lower curve part b) Upper curve part

The *MSE* (equation 3.4) for our example is the summation of the squared distances for the all points from the polygonal approximating straight lines divided by the original contour length

$$MSE = \frac{1}{43}(0.5 + 20.2 + 14.3302 + 96.0718) = 3.0489$$

The variance, often denoted by σ^2. The variance is a measure of how spread out a distribution is. It is computed as the average squared deviation of each number from its mean. Numerically, the variance equals the average of the several squared deviations from the mean.

In our example the mean value (μ) of the original contour is 72.6395.

We calculate the variance using equation (3.6).

In our example the N is equals to (number of rows * number of columns) = (2*43) = 86.

The variance of the input of the original contour is

$$\sigma^2 = \frac{\sum_{n=1}^{2}\sum_{m=1}^{43}(CC_{nm} - 72.6395)^2}{86} = 569.6956$$

where CC_{nm} is the input of the original contour.

Finally the *SNR* using equation (3.5) for our example is

$$SNR = -10 * \log_{10}\left(\frac{3.0489}{569.6956}\right) = 22.7150$$

Hint: Using Matlab we can simply make compression for the extracted contour by calling the file AproxRamer as

```
>> TA = AproxRamer ( CC, 4 );
```

where: CC is sequence of the input for the contour and 4 is the threshold value.

3.3 Triangle method

Fits the edge points with a sequence of line segments (i.e., contour is represented as a polygon).There are several algorithms available for determining the number and location of the vertices and also to compute the polygonal approximation of a contour. The Triangle method is based on the polygonal approximation scheme. The threshold is depends on the length between each two points of the triangle If the ratio value between the height and the base line of the triangle is exceeds the threshold value then the starting point is shifted to the second point of the triangle and open a new segment. The starting point is stored as vertex of an edge of the approximating contour. Otherwise; the starting point is shifted to the ending point and open a new segment. The starting point is stored as vertex of an edge of the approximating contour. Finally draw the lines between the vertices to obtain the approximating contour [37] and [38].
To analyze the Triangle method let us take the following example.
Consider the closed contour (i.e the first and last points of the contour are equals) as shown in Fig. 3.30. The contour is represented by (y-x) Cartesian Co-ordinates. The sequence of the contour (*CC*) is shown in Table 3.7.

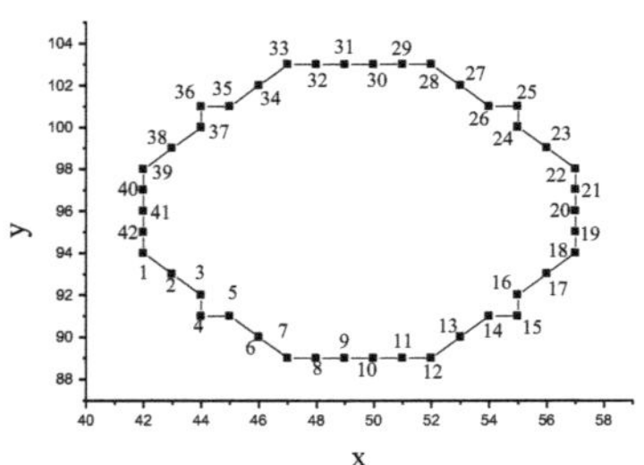

Fig. 3.30. The original contour

Table 3.7

PN	1	2	3	4	5	6	7	8	9	10	11	12	13	14
x	42	43	44	44	45	46	47	48	49	50	51	52	53	54
y	94	93	92	91	91	90	89	89	89	89	89	89	90	91
PN	15	16	17	18	19	20	21	22	23	24	25	26	27	28
x	55	55	56	57	57	57	57	57	56	55	55	54	53	52
y	91	92	93	94	95	96	97	98	99	100	101	101	102	103
PN	29	30	31	32	33	34	35	36	37	38	39	40	41	42
x	51	50	49	48	47	46	45	44	44	43	42	42	42	42
y	103	103	103	103	103	102	101	101	100	99	98	97	96	95
PN	43													
x	42													
y	94													

Where: PN is the point number (order) in the contour sequence.

First, the two input parameters should be known. First one is the length (N) between the each two points in the Triangle (our choice is 6). Second parameter is the threshold value (let us say 0.5).

The analysis steps are:

1. Determine the start point SP (x_1, y_1) by looking for the minimum value of x in the lower left side of the contour (42, 94) (see Fig. 3.31).

2. Determine the second point B (x_2, y_2) by adding N to the starting point number (i.e. the point number 7 (47, 89)) as shown in the Fig. 3.31).

3. Determine the end point EP (x_3, y_3) by adding N to the point B (i.e. the point number 13 (53, 90)) as shown in the Fig. 3.31).

4. Store the starting point number in the sequence of the contour as a vertex (AC = [1]).

5. Draw the straight line between the starting and B points (see Fig. 3.31).

6. Draw the straight line between the starting and ending points (see Fig. 3.31).

7. Draw the straight line between the ending and B points (see Fig. 3.31).

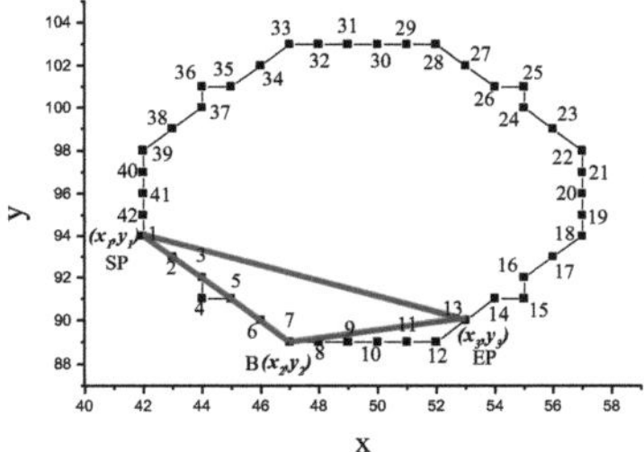

Fig. 3.31. The Starting, B and ending points of the triangle

8. Calculate the distance between the starting point (42, 94) and the B point (47, 89) using the following equation (see Fig. 3.32)

$$SB = \sqrt{(y_2 - y_1)^2 + (x_2 - x_1)^2} = \sqrt{(89-94)^2 + (47-42)^2} = 7.0711$$

The number of operations is 13.

9. Calculate the distance between the starting point (42, 94) and the ending point (53, 90) using the following equation (see Fig. 3.32)

$$SE = \sqrt{(y_3 - y_1)^2 + (x_3 - x_1)^2} = \sqrt{(90-94)^2 + (53-42)^2} = 11.7047$$

The number of operations is 19.

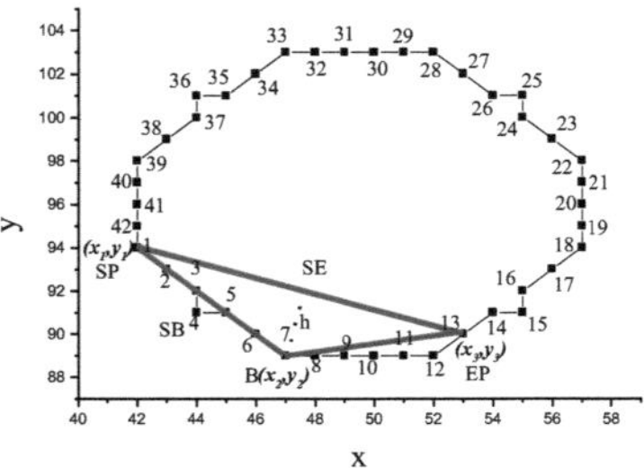

Fig. 3.32. The *h*, *SB* and *SE* distances of the triangle

10. Calculate the slope m_1 for the line *SE* using the following equation

$$m_1 = \frac{y_3 - y_1}{x_3 - x_1} = \frac{90-94}{53-42} = \frac{-4}{11} = -0.3636$$

11. Calculate the slope m_2 for the line *SB* using the following equation

$$m_2 = \frac{y_2 - y_1}{x_2 - x_1} = \frac{89-94}{47-42} = \frac{-5}{5} = -1$$

12. Calculate the angle (ϕ) between the lines *SB* and *SE* using the following equation (3.7)

$$\varphi = \tan^{-1}\left(\frac{-0.3636-(-1)}{1+(-0.3636)*(-1)}\right) = \tan^{-1}(0.4667) = 0.4367$$

13. Calculate the absolute value of the height of the triangle (*h*) using the following equation (see Fig. 3.32)

$$h = \left|SB * \sin(\tan^{-1}(\phi))\right| \tag{3.8}$$

$$h = |7.0711*(0.4230| = 2.99$$

The number of operations is 30.

14. Calculate the ratio between the h and SE.

In the example the ratio is $\frac{h}{SE} = \frac{2.99}{11.7047} = 0.2555$

15. If the ratio value exceeds the threshold (0.5), the B point is stored as a vertex of an edge of the approximating polygon. The SP point is shifted to the B point of the triangle and then a new segment is drawn. Otherwise; the end point EP of the triangle is stored and the start point SP is shifted to the end point EP, then a new segment is drawn.

In our example the ratio value is not exceeds the threshold and hence the number of EP is stored as a vertex of an edge of the approximating polygon as a vertex of an edge of the approximating polygon. So the approximating contour vector is AC= [1 13]. Now check whether the new end point is exceeds the length of the input contour or not. The number of the new EP is 25 and it is not exceeds the length of the input contour. Put the SP equals to the EP (53, 90) (see Fig. 3.33). The B point number is equals to point number 19 (57, 95) by adding N to the starting point number (see Fig. 3.33). The EP is equals to the point number 25 (55, 101) by adding N to the B point number (see Fig. 3.33).

16. This step is to show what happened in case of the ratio value is exceeds the threshold value. If the ratio is exceeds the threshed value (0.5) then there is no edge point between the starting and ending points of the triangle will be stored. So the starting point is shifted to the second point (B) number (7) and a new segment is opened. This step is replaced by the step 15 only in case of the ratio value is exceeds the threshold value. The vector of the approximating contour is AC = [1 7].

17. Let us back to the end of step 15. Draw the straight line between the starting and B points (see Fig. 3.33).

Now for the new triangle the SP is (53, 90), B point is (57, 95) and EP is (55, 101)

18. Draw the straight line between the starting and B points (see Fig. 3.33).

19. Draw the straight line between the starting and ending points (see Fig. 3.33).

20. Draw the straight line between the ending and B points (see Fig. 3.33).

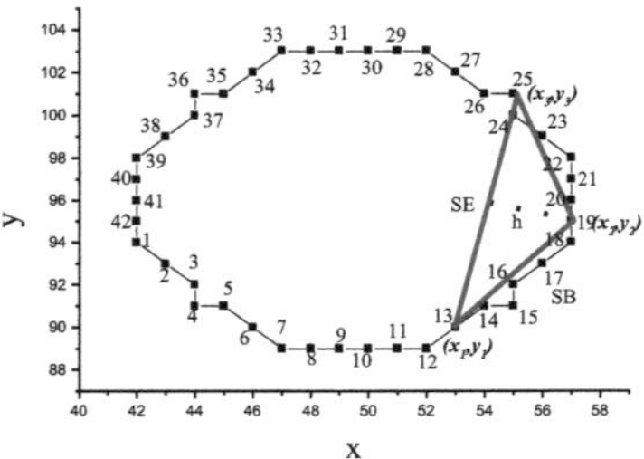

Fig. 3.33. The *h*, *SB* and *SE* distances of the second triangle.

21. Repeat the steps from 8 to 14 to calculate the following values:

SB = 6.4031, *SE* = 11.1803, m_1 =5.5, m_2 =1.25, $\phi = 0.4949$ and *h* = 3.00411. The number of operations is 61.

The ratio value is 0.2720 and it is not exceeds the threshold (0.5) and hence the number of EP is stored as a vertex of an edge of the approximating polygon. So the approximating contour vector is AC= [1 13 25]. Now check whether the new end point is exceeds the length of the input contour or not. The number of the new EP is 38 and it is not exceeds the length of the input contour. Put the SP equals to the EP (55, 101) (see Fig. 3.34). The B point number is equals to point number 31 (49, 103) by adding N to the starting point number (see Fig. 3.34). The EP is equals to the point number 37 (44, 100) by adding N to the B point number (see Fig. 3.34).
Now for the new triangle the SP is (55, 101), B point is (49, 103) and EP is (44, 100).

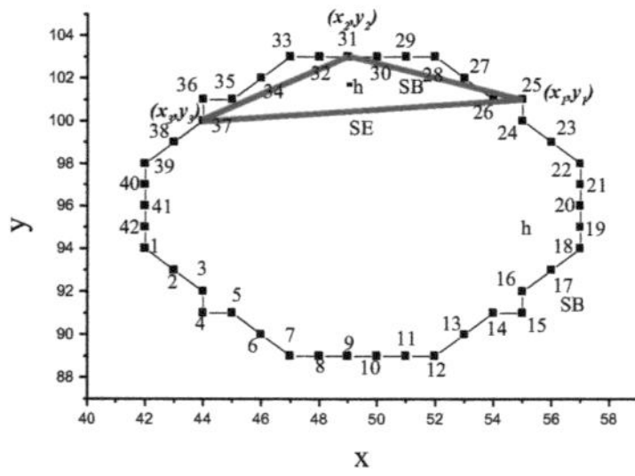

Fig. 3.34. The *h*, *SB* and *SE* distances of the third triangle

22. Repeat the steps from 8 to 14 to calculate the following values:

$SB = 6.3246$, $SE = 11.0454$, m_1 =0.0909, m_2 =-0.3333, $\phi = 0.4124$ and $h = 2.5350$. The number of operations is 92.

The ratio value is 0.2295 and it is not exceeds the threshold (0.5) and hence the number of EP is stored. So the approximating contour vector is AC= [1 13 25 37]. Now check whether the new end point is exceeds the length of the input contour or not. The number of the new EP is 49 and it is exceeds the length of the input contour and hence break the algorithm. To close the contour the first point is restored as the last point and finally the AC vector is [1 13 25 37 1]. The number of operations is 94.

The original contour and the approximating contour (AC) is shown in Fig. 3.35.

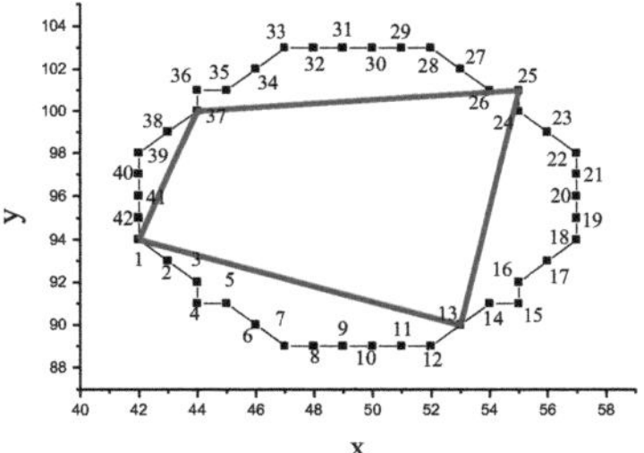

Fig. 3.35 The Original and approximating contours

As we can see the original contour contains 43 points while the approximating contour contains only 5 points. This ratio between these values is $(\frac{5}{43}) = 0.1163$ which is means that the 0.1163 of the contour information is reconstructed the original contour while the rest is rejected.

The compression ratio (equation 3.3) for our example is $CR = \frac{(43-5)}{43} * 100\% = 88.37\%$

In our example the approximating contour has four straight lines:

The straight line l_1 between the points number 1 and 13, the straight line l_2 between the points 13 and 25, the straight line l_3 between the point's number 25 and 37 and the straight line l_4 between the points number 37 and 1.
The calculation procedure has the same steps as in Ramer method. After long calculation the *MSE* (equation 3.4) is 3.3468.

The *SNR* (equation 3.5) is equals to 22.3102 decibel.

To make comparison between the Triangle and Ramer methods the Table 3.8 shows their parameters values.

Table 3.8

	MSE	*SNR*	*CR* [%]	NO
Ramer method	3.0489	22.7150	88.37	803
Triangle method	3.3468	22.3102	88.37	94

NO – number of the operations.

The Table 3.8 show the *MSE* and *SNR* values for the Ramer method is very closed to the values for the Triangle method. The number of operations in Ramer method is too large when compared with the number of operations in Triangle method. For our example (small contour only 43 points), the Triangle method is more than eight (8) times faster than the Ramer method.
The large number of operations in Ramer method is because after test the points of the curve segment to find the maximum distance from the straight line, the curve segment is broken into two new segments at the curve point most distant from the straight-line segment. Then the points in each these new segments are tested once more to find the maximum distance from the new straight line. This segmentation is repeated until each curve segment can be approximated by a straight-line segment through its endpoints. And that is taking a lot of operations. In our example to store the first an edge point as a vertex of the approximating contour the number of operations is 260.
In Triangle method to store the first an edge point as a vertex of the approximating contour we need only to find the distance between SP and B points and the distance between the SP and the EP points of the triangle and also the angle between them. Finally we need to find the height of the triangle. This means that in our example to store the first point as a vertex of an edge of the approximating contour we need only 31 operations.
Fig. 3.36 shows the original contour and the approximating contour using the Ramer and Triangle methods.
Hint: Using Matlab we can call the file AproxTrihb as
>> TA = AproxTrihb (CC, 0.5, 6);
where: hb is the ratio between the height and length of the triangle; 0.5 is the threshold value and 6 is the length between two points in the triangle.

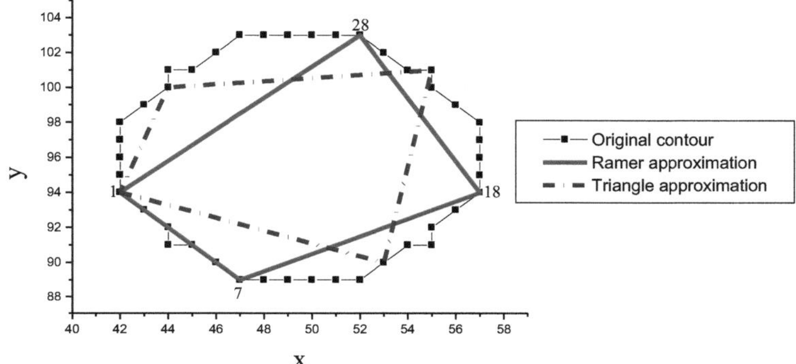

Fig. 3.36 The Original, Ramer approximating and Triangle approximating contours

Hint: Using Matlab we can simply make compression for the extracted contour by calling the file AproxTrihb as

```
>> TA = AproxTrihb ( CC, 0.5, 6 );
```

where: hb means that we compared the ratio between the height and length of the triangle with the threshold value; CC is sequence of the input for the contour; 0.5 is the threshold value and 6 is the length between two points in the triangle.

3.4 Trapezoid method

The proposed algorithm belongs to a family of polygonal methods of approximation. The idea of this method consists in segmentation of the contour points to get trapezoid shapes (points of SP, B, C, and EP). The first and last points of each segment are called starting point (SP) and ending point (EP) respectively. The fit criterion of the Trapezoid method [9]; is the ratio between the perpendicular distance (dB) from point B to the line (SP – C) to the perpendicular distance (dC) from point C to the line (B – EP), as illustrated in Fig. 3.37, and is defined by equation (3.9) [39], [40] and [41].

$$(dB / dC) < th \tag{3.9}$$

where th is given threshold value.

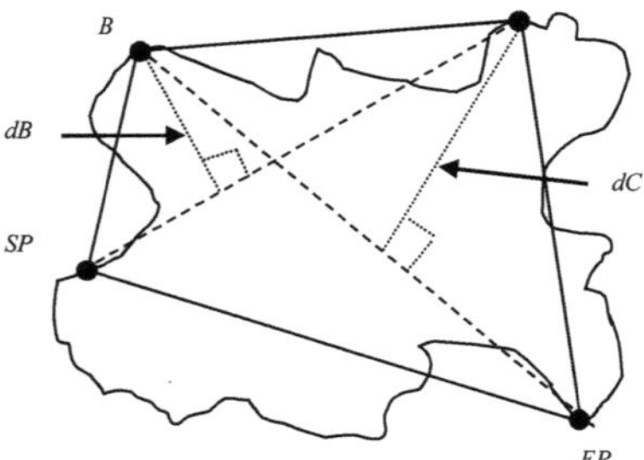

Fig. 3.37 Fit criterion of the Trapezoid method algorithm

Trigonometric formula is used to calculate these values. If equation (3.9) is not valid the second, third and ending points are stored and the SP is shifted to the EP of the trapezoid, then a new segment is drawn. Otherwise the third and ending points of the trapezoid are stored and the SP is shifted to the EP; then a new segment is drawn. The vertices of an edge of the approximating polygon are determined by these stored points. The flowchart showing the idea of Trapezoid method is shown in Fig. 3.38; where VA, CC, L_{CC}, and f are sequence of indices of the final vertices, sequence of the contour input, the input contour length, and the length between each two points of the trapezoid respectively. Only the co-ordinates of the first and last points of the contour segment are stored.

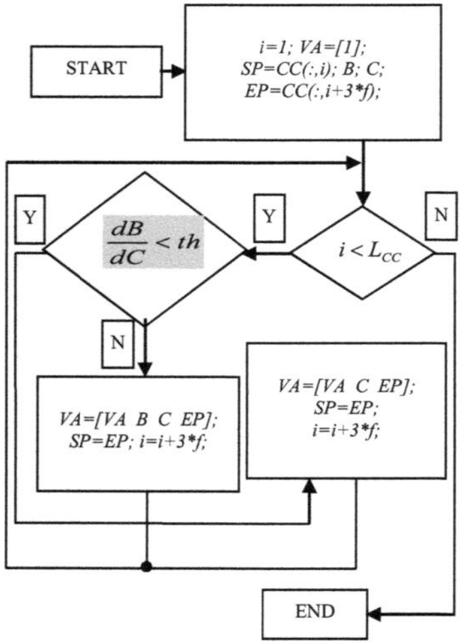

Fig. 3.38 Flowchart of the Trapezoid method algorithm

The analysed criterion of the algorithm can be modified depending on methods of contour representation. The contour uses chain coding schemes of contour representations to determine all possible connections for both 8-connectivity and 4-connectivity schemes such as Cartesian representation, or polar representation, or generalized [26].

The analysed algorithms use new method of contour extraction (SSPCE) with 3x3 pixels window structure. By using the central pixel the object contours is extracted and the all possible edge direction is found which connects the central pixel with one of the remaining pixels surrounding it.

The compression ratio (CR), mean square error (MSE), and signal-noise-ratio (SNR) between the original & compressed contours can be calculated using equations (3.3, 3.4, and 3.5) respectively.

```
>> AC = AproxTrapezoid (CC,th,f)
```

where: AC is the approximated contour and AproxTrapezoid means that the contour is compressed using Trapezoid method. Distances ratios are compared with the threshold (th) value; CC is sequence of the input for the contour; f is the length between two points in the trapezoid shape.

3.5 Adaptive-Triangle Method

This section presents a suggestion of a new algorithm for contour compression of black-white images in time domain; based on polygonal approximation. The idea of this algorithm [42] based on choosing two points (S_p, E_p) on the contour curve as in (Fig. 3.39). S_p is point number 1 on the curve, E_p is point number 12 on the same curve. Now; the straight line that connecting (S_p,E_p) is used to compute the maximum perpendicular distance (d_{max}) from this line to contour curve points from S_p to E_p to as shown in (Fig. 3.39), compare this distance (d_{max}) with given threshold (th_1) by the equation (3.10)

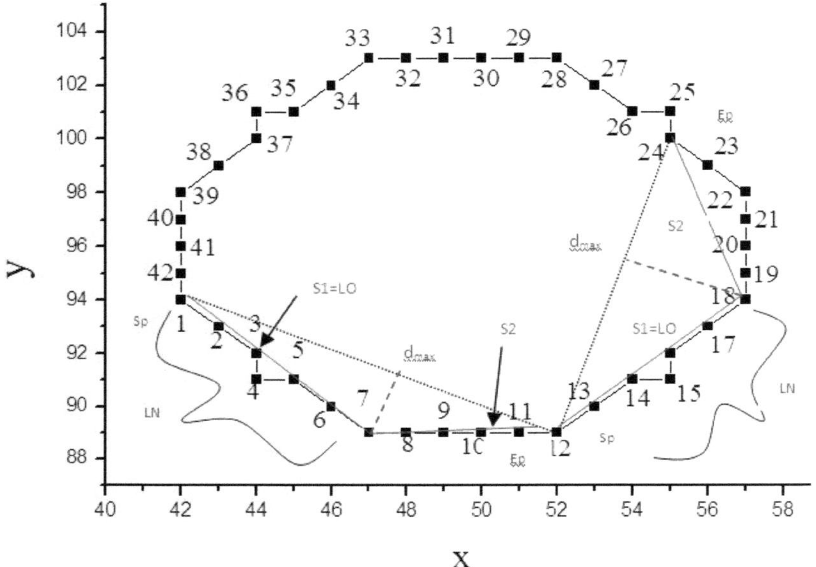

Fig. 3.39 Fit criterion of the Adaptive-Triangle method algorithm

$$d_{\max} > th_1 \tag{3.10}$$

There are two possible cases:

1. If the distance (d_{max}) is larger than (th_1) and equation (3.10) comes true, we will open two new line segments(S_1 , S_2), S_1 is the direct line that connect S_p to d_{max} point, and S_2 the direct line that connect d_{max} point to E_p . Now, compare the length of S_1 , S_2 lines:

- If ($S_1 > S_2$) we will compare the ratio between (Lo :direct length) of largest segment , with the (L_N: real length) of the same segment with a given threshold (th_2) according to equation (3.11)

$$(L_o / L_N) < th_2 \quad (3.11)$$

If the condition of equation (3.11) comes true, we choose two additional points on the (S_1 , S_2) lines as a vertex points of compressed contour, so the vertex points will be:

Vertex Points= [S_p $S_1/2$ $S_2/2$ E_p], and start point S_p is shifted to the end point E_p and open new segment, as shown in (Fig. 3.39).

- Otherwise we will store Vertex Points = [S_p $S_{largest}/2$ E_p] as a vertex points of compressed contour, and start point S_p shifted to end point E_p and open new segment, as hown in (Fig. 3.39).

2. Second case, if condition of equation (3.10) fails, i.e.

$$d_{\max} < th_1 \quad (3.12)$$

We will store (S_p,E_p) as a vertex points , and shift S_p to E_p and open new segment, as shown in (Fig. 3.39). The flowchart of this algorithm shown in (Fig. 3.40).

```
>> AC = AproxTriangle_Adaptive(CC,th1,f,th2);
```

% where: AC is the approximated contour and AproxTriangle_Adaptive means that the contour is compressed using Adaptive-Triangle method. Distances ratios are compared with two thresholds (th1 & th2) values; CC is sequence of the input for the contour; f is the length between two points in the triangle shape.

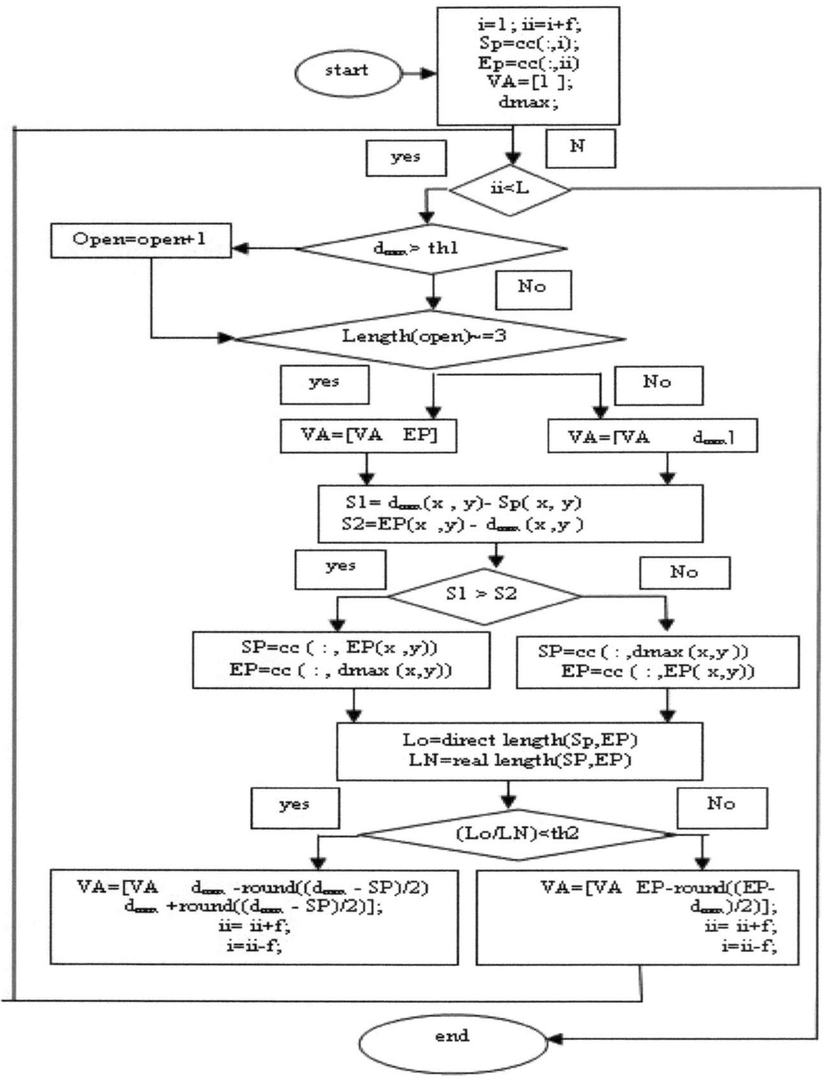

Fig. 3.40 Flowchart of the Trapezoid method algorithm

The compression ratio (CR), mean square error (MSE), and signal-noise-ratio (SNR) between the original & compressed contours can be calculated using equations (3.3, 3.4, and 3.5) respectively.

3.6 Experimental Results

>> quit = AproxRamer(CC,3.17); % contour approximation using Ramer algorithm where 3.17 is the threshold value.

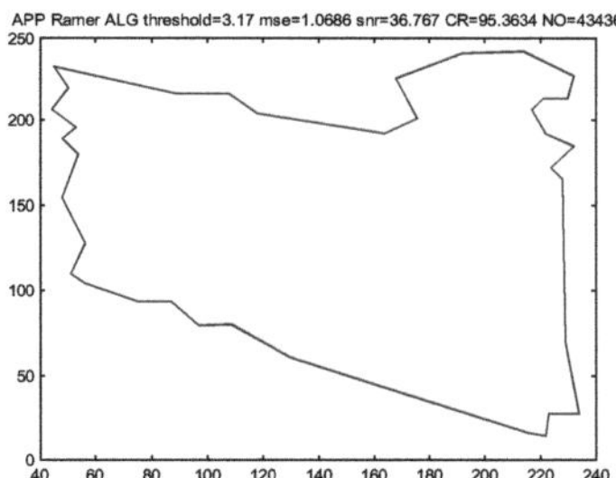

Fig. 3.41 Contour approximation using Ramer method

>> quit = AproxTangent(CC,33,20); % contour approximation using Tangent algorithm where 33 is the threshold value and 20 is the maximum width of the segment.

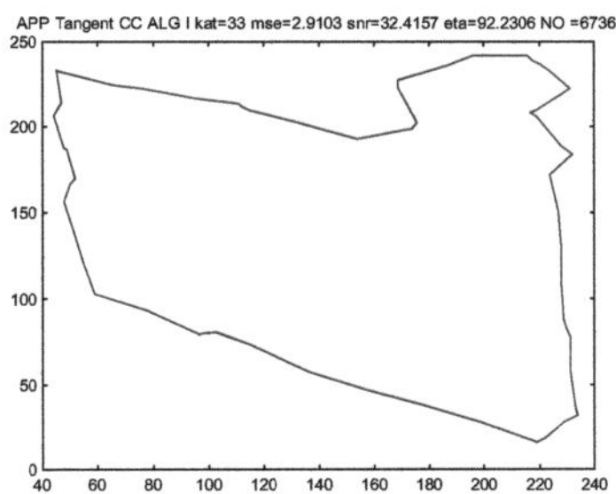

Fig. 3.42 Contour approximation using Tangent method

>> quit = AproxTrihb(CC,0.33,9); % contour approximation using Triangle algorithm where 0.33 is the threshold value and 9 is the distance between any two points in the triangle shape of the segment.

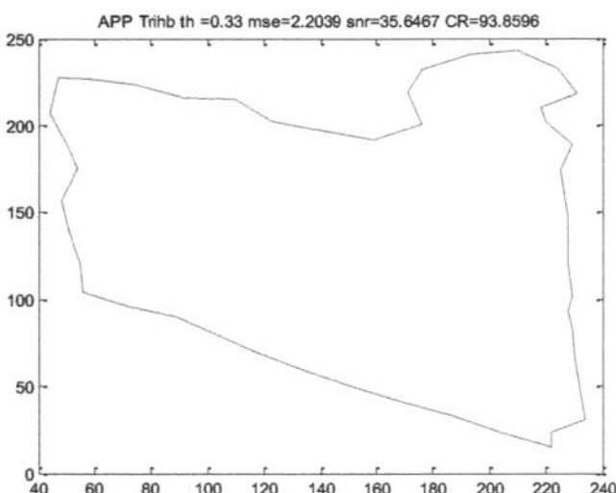

Fig. 3.43 Contour approximation using Triangle method

quit = AproxTriangle_Adaptive(CC,9,15,1.7); % contour approximation using Adaptive-Triangle algorithm where 9 is the first threshold value and 15 is the distance between any two points in the triangle shape of the segment and 1.7 is the second threshold value.

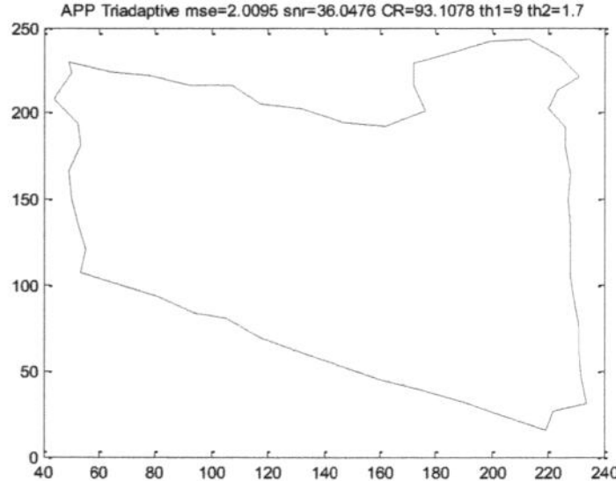

Fig. 3.44 Contour approximation using Adaptive-Triangle method

where: mse is mean square error; snr is signal-to-noise ratio; CR is compression ratio; kat is the angle between starting & closing segments & NO is the number of operations.

Comparison between these methods of different contour compression are shown from Fig. 3.45 to Fig. 3.50 using CR, MSE, SNR & consumed time criterions (related results are in Tables 3.9 & 3.10.

Table 3.9 Results of Compression for 'Libya Map' Contour using (Ramer, Trapezoid, Triangle Adaptive) Methods

	Compression Method	*CR*%	*SNR* (dB)	*Time* 'ms'
Libya map	Ramer	79.9127	44.9098	69.2
		85.5895	45.0827	61.11
		90.1019	43.108	56.7
		98.2533	28.2568	52.26
	Triangle Adaptive	79.7671	45.4323	47.7
		85.1528	43.1515	46
		90.393	39.032	45.9
		97.8166	25.6039	40.09
	Trapezoid	79.6215	42.7478	39
		85.444	42.1536	36.7
		90.1019	36.147	36.08
		97.8166	22.087	34.1

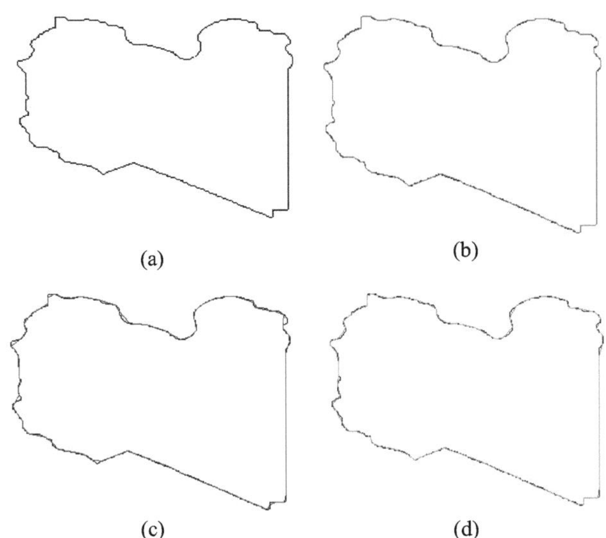

Fig. 3.45 (a) Original Contour of 'Libya Map', (b,c,d) Compressed Contour of 'Libya Map' at *CR*=85% using (Ramer, Trapezoid, Triangle) methods respectively

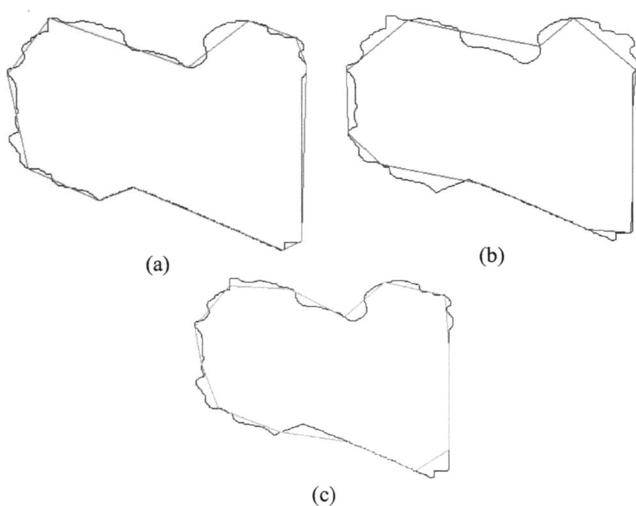

Fig.3.46 (a,b,c) Compressed Contour of 'Libya Map' at CR=98% using (Ramer, Trapezoid, Triangle) methods respectively

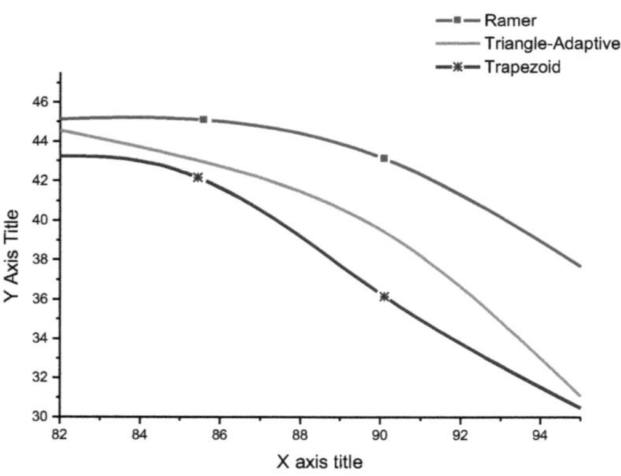

Fig. 3.47 *SNR* for Ramer, Trapezoid, and Triangle Adaptive compression algorithms of 'Libya map' as a function of *CR*

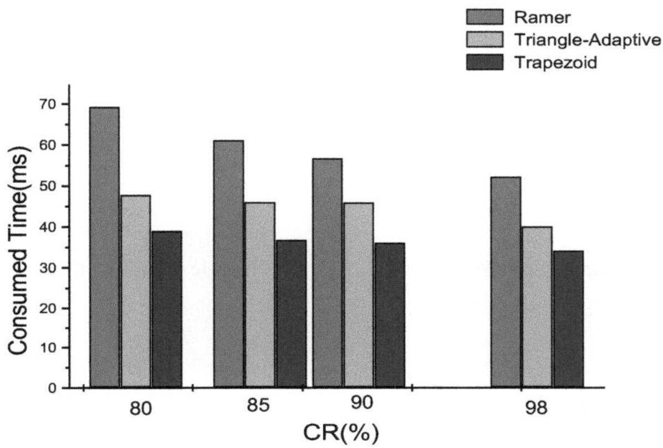

Fig. 3.48 Consumed time versus CR

Table 3.10 Results of Compression of "Tools" Contour using (Ramer, Trapezoid, Triangle Adaptive) Methods

	Compression Method	*CR*%	*SNR* (dB)	*Time* 'ms'
Tools	Ramer	80.6133	45.9624	333
		85.0789	48.916	307.01
		90.3156	46.436	308.19
	Triangle Adaptive	79.7704	49.463	307.78
		85.0789	48.4371	296.98
		90.1004	44.4551	278.3
	Trapezoid	79.8422	45.106	277.9
		85.3659	46.5629	283.1
		90.0287	41.3202	276.7

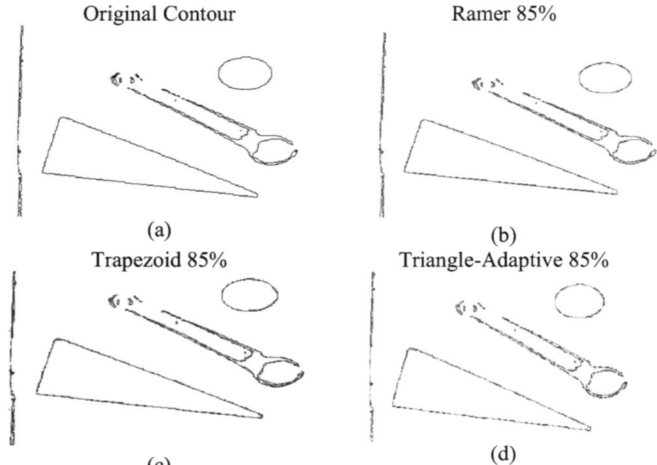

Fig. 3.49 (a) Original Contour , (b,c,d) Compressed Contour of Cameraman at *CR*=85% using (Ramer, Trapezoid, Triangle) methods respectively

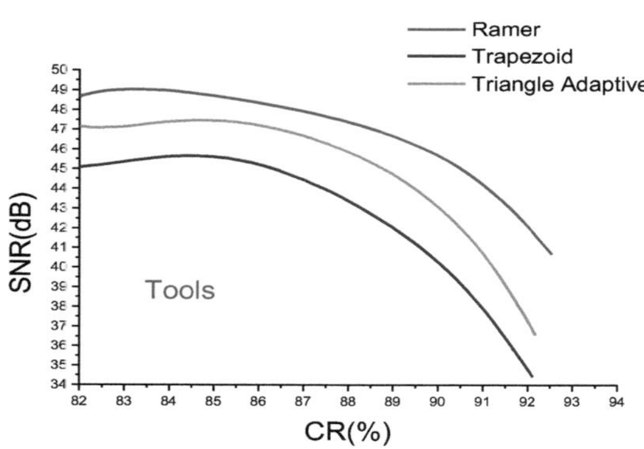

Fig. 3.50 Relation between *CR*(%), SNR(dB) for Extracted Contours of 'Tools'

To visualize the experimental results a Roses contour is selected as shown in Fig. 3.51 (related results are shown in Fig. 3.52 & Fig. 3.53 and Table 3.11). First contour is Roses and it consists of (4596) points;

Fig. 3.51 Original Roses contour

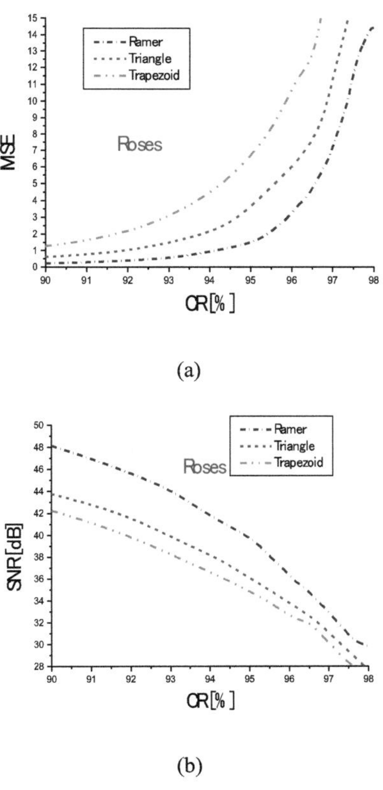

(a)

(b)

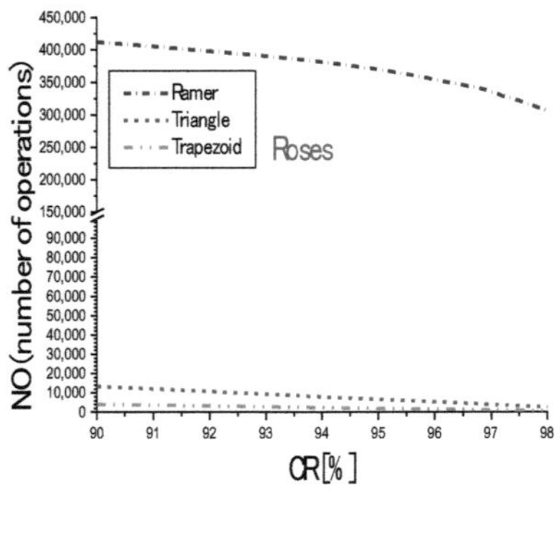

(c)

Fig. 3.52 Roses contour approximation: (a) MSE versus CR, (b) *SNR* versus CR, and (c) NO versus CR

Table 3.11

Method	*MSE*	*SNR*	*CR* [%]	NO
a) Ramer	3.25	37.81	96.02	353218
b) Triangle	9.00	33.41	96.06	5205
c) Trapezoid	11.86	32.19	96.02	1658

a)

b)

c)

Fig.3.53 (a,b,c) Compressed Contour of 'Roses' at *CR*=96% using (Ramer, Trapezoid, Triangle) methods respectively

Chapter Four
Contour Compression Methods in Spectral Domain

4.1 Overview

The transform coding method compresses image data by representing the original signal with a small number of transform coefficients. It exploits the fact that for typical images a large amount of signal energy is concentrated in a small number of coefficients. The goal of transform coding is to minimize the number of retained transform coefficients while keeping distortion at an acceptable level. Transform coding is an integral part of one of the most widely known standards for lossy image compression, the JPEG (Joint Photographic Experts Group) standard.

In this chapter we look at a number of different transforms, including the popular Walsh transform [48], [49], [50], [51] and [52]. There exists a class of transforms, based on the integration of Haar functions, known as Periodic Haar Piecewise-Linear (PHL) transform [53]. The main purpose of this chapter is to make comparisons for contour compression between time domain algorithms (Ramer & Triangle Methods) and transform domain algorithms (DCT [54] and Haar [56]).

To present the results, the mean square error and signal-to-noise ratio criterions versus compression ratio are used. Computational times of analysed procedures are estimated based on a number of numerical operations.

The extraction steps are done then the compression steps follow. Fig. 4.1 shows two stages. The first stage is the extracted contour; while the second one is the compressed contour. The input of the first stage is two-dimensional grey-scale image and the extracted contour is the output of that stage. The input of the second stage is the contour signal (X(n) and Y(n)); and the output is the approximated contour (*X(n)* and *Y(n)*).

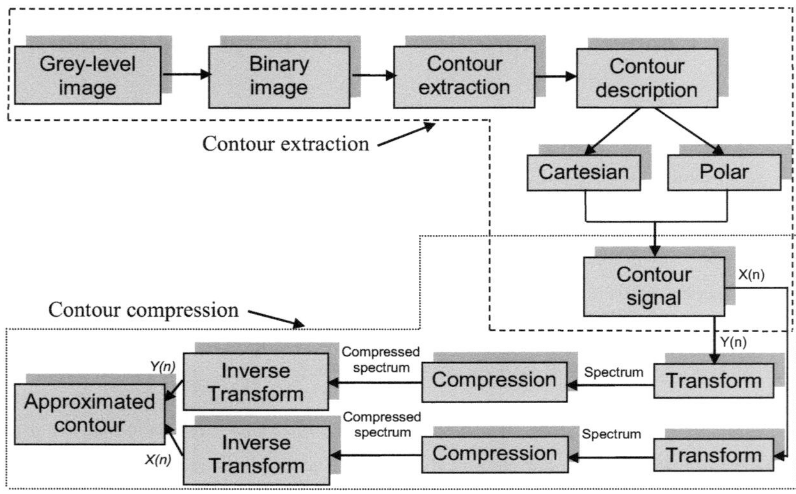

Fig 4.1 Block diagram for contour extraction and compression in transform domain

4.2 Discrete Cosine Transform (DCT)

The Discrete Cosine Transform (DCT) helps separate the image into parts (or spectral sub-bands) of differing importance (with respect to the image's visual quality). The DCT is similar to the Discrete Fourier Transform (DFT): it transforms a signal or image from the spatial domain to the frequency domain. DCT is similar to the Fast Fourier Transform (FFT), but can approximate lines well with fewer coefficients.

The DCT gets its name from the fact that the rows of the transform matrix C are obtained as a function of cosines.

$$[C]_{i,j} = \begin{cases} \sqrt{\frac{1}{N}}\cos\frac{(2j+1)i\pi}{2N} & for \quad i=0, j=0,1,\ldots,N-1 \\ \sqrt{\frac{2}{N}}\cos\frac{(2j+1)i\pi}{2N} & for \quad i=1,2,\ldots,N-1, j=0,1,\ldots,N-1 \end{cases} \tag{4.1}$$

It is computationally easier to implement and more efficient to regard the DCT as a set of basis functions which given a known input array size (8 x 8) can be pre-computed and stored. This involves simply computing values for a convolution mask (8 x8 window) that get applied (sum values x pixel the window overlap with image apply window across all rows/columns of image). The values are simply calculated from the DCT formula. The 64 (8 x 8) DCT basis functions are illustrated in Fig. 4.2.

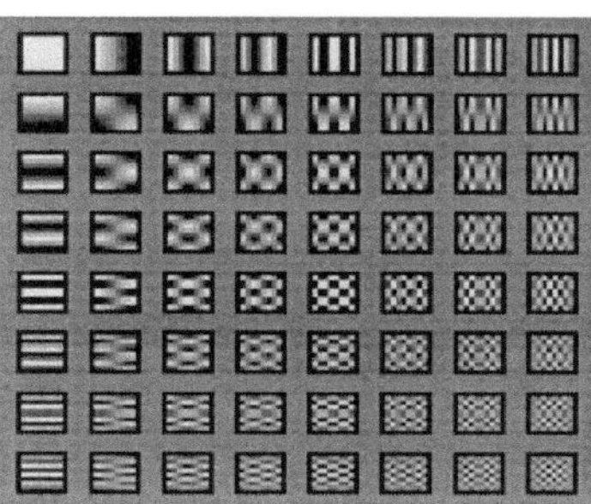

Fig. 4.2 The basis images for the DCT transform

The 2-D DCT block calculates the two-dimensional discrete cosine transform of the input signal. The equation for the two-dimensional DCT is

$$F(m,n)=\frac{2}{\sqrt{MN}}C(m)C(n)\sum_{x=0}^{M-1}\sum_{y=0}^{N-1}f(x,y)\cos\frac{(2x+1)m\pi}{2M}\cos\frac{(2y+1)n\pi}{2N} \qquad (4.2)$$

where $C(m),C(n)=\frac{1}{\sqrt{2}}$ *for* $m,n=0$ and $C(m),C(n)=1$ *otherwise.*

DCT is similar to the Fast Fourier Transform (FFT), but can approximate lines well with fewer coefficients. To obtain the output array of DCT coefficients, apply 1D DCT (vertically) columns and apply 1D DCT (horizontally) to resultant vertical DCT above or alternately horizontal to vertical.

Most software implementations use fixed point arithmetic. Some fast implementations approximate coefficients so all multiplies are shifts and adds. All DCT multiplications are real. This lowers the number of required multiplications, as compared to the discrete Fourier transform. The DCT input is an 8 by 8 array of integers. This array contains each pixel's gray scale level; 8 bit pixels have levels from 0 to 255. The output array of DCT coefficients contains integers; these can range from -1024 to 1023. For most images, much of the signal energy lies at low frequencies; these appear in the upper left corner of the DCT. The lower right values represent higher frequencies, and are often small - small enough to be neglected with little visible distortion.

4.3 Contour Compression in Transform Domain

To do the transform for contour the length of the contour should be equals to 2^n (this value should be greater than or equals to the length of the contour i.e. 798); where n is an integer number.

In our example the length of CC is 798 and then the n is 10 where 2^10 = 1024. This means that the points from 799 to 1024 are zeros.

>> CCM = zeros(2,1024); % matrix size is (2 x 1024), all of them are zeros.

>> CCM(1:2,1:length(CC))=CC(1:2,1:length(CC)); % Paste the CC contour into the CCM matrix.

>> FDCTX = dct(CCM(1,:)); % Transform the X points into spectral domain using Discrete Cosine Transform (DCT).

>> FDCTY = dct(CCM(2,:));

Transform the Y points into spectral domain using Discrete Cosine Transform (DCT).

Here in spectral domain can make the thresholding. To do a certain threshold we should plot the spectral of the contour (for X and Y respectively).

>> figure; plot(FDCTX) % Fig.4.3

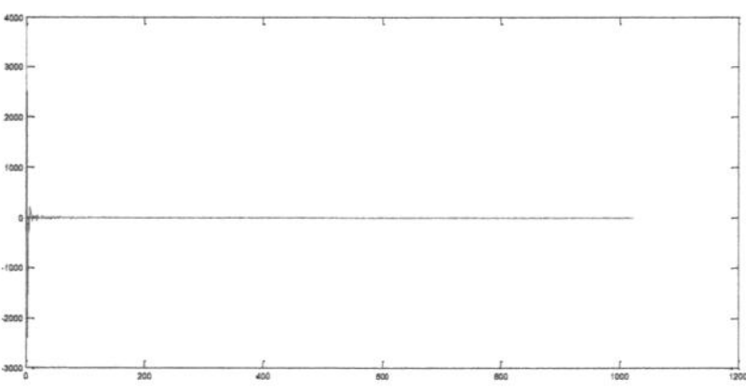

Fig. 4.3 X coefficients plot using DCT

>> figure; plot(FDCTY) % Fig. 4.4

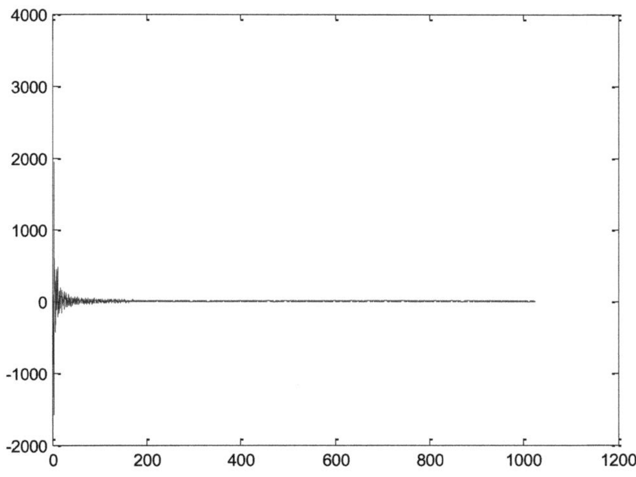

Fig. 4.4 Y coefficients plot using DCT

>> T = [FDCTX; FDCTY];

>> T = (abs(T)>.6).*T; % This means that we leave any coefficient has absolute value greater than 0.6 and the others put it zeros.

>> IDCTX = idct(T(1,:)); % Inverse Transform of the X points using Discrete Cosine Transform (DCT).

>> IDCTY = idct(T(2,:)); % Inverse Transform of the Y points using Discrete Cosine Transform (DCT).

>> figure; plot(CC(1,:)); % Plot the X points of the contour (Fig. 4.5).

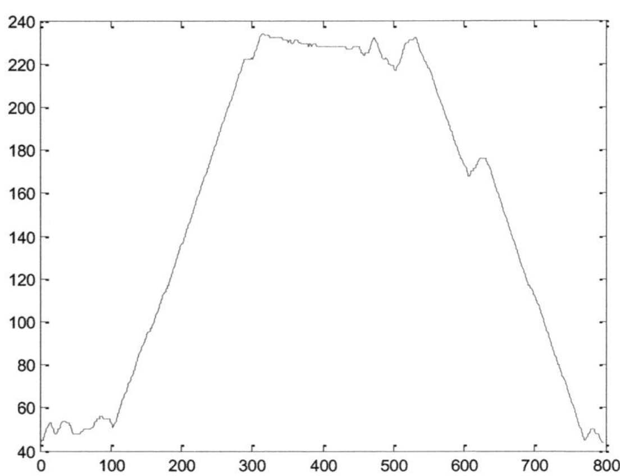

Fig. 4.5 X plot of original contour in spatial domain

>> figure; plot(IDCTX(1,1:length(CC)));%Plot the reconstruction of the X points (Fig. 4.6).

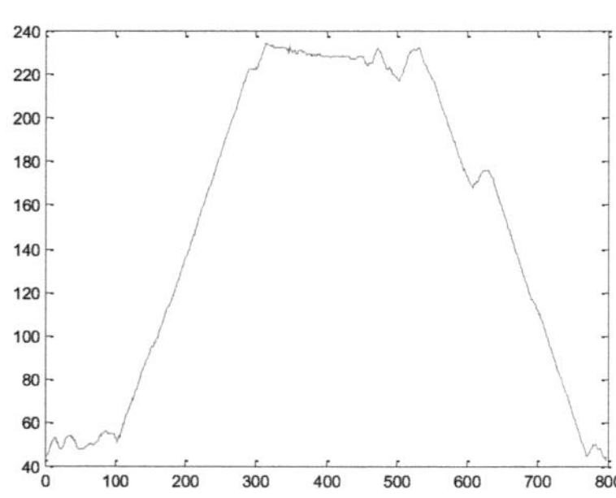

Fig. 4.6 X plot of reconstructed contour in spatial domain

>> figure; plot(CC(2,:)); % Plot the Y points of the contour (4.7).

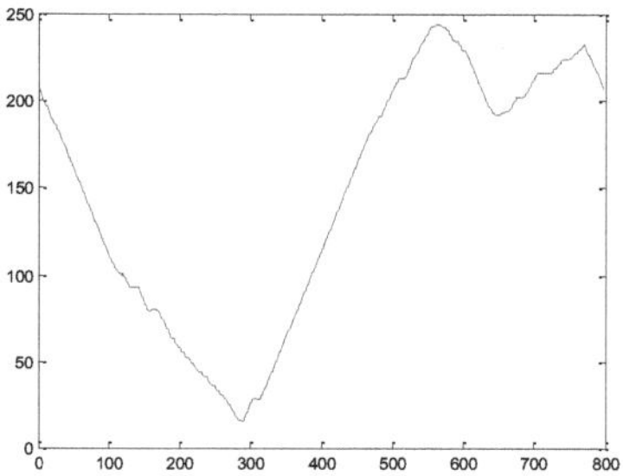

Fig. 4.7 Y plot of original contour in spatial domain

>> figure; plot(IDCTY(1,1:length(CC)));% Plot the reconstruction of the Y points (Fig. 4.8).

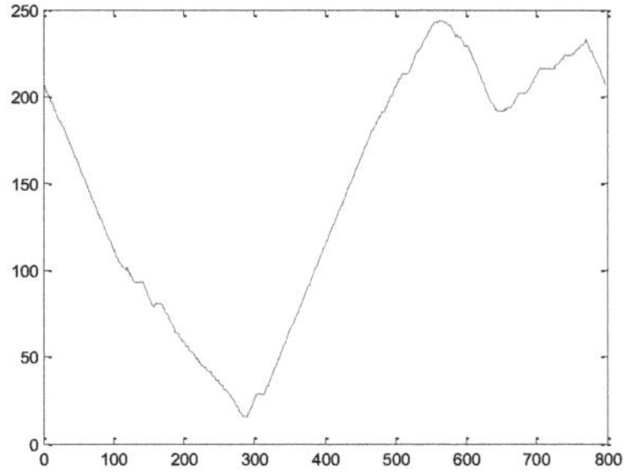

Fig. 4.8 Y plot of reconstructed contour in spatial domain

>> figure; plot(IDCTX(1,1:length(CC)), IDCTY(1,1:length(CC))); % Plot the reconstruction of the contour (Fig.4.9).

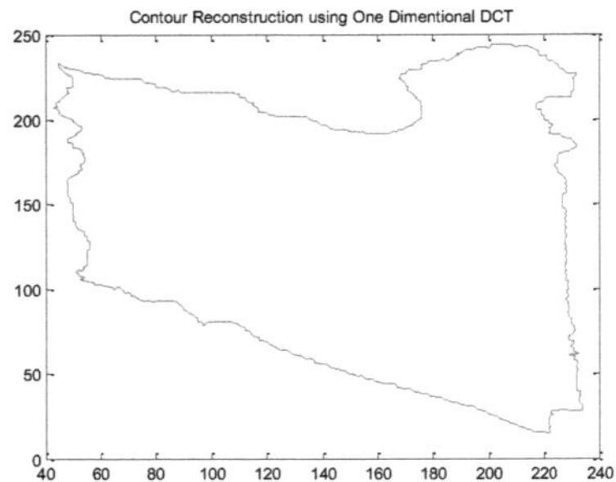

Fig. 4.9 The reconstructed contour in spatial domain

Fig. 4.10 shows the DCT transform is performed for each of X(n) and Y(n) for the 'Italy map' contour; then the threshold sampling is done and finally the IDCT is applied for each X and Y of DCT to reconstruct the Italy map approximated contour (the related results are shown in Table 4.1).

Table 4.1 Results of approximation for the Italy contour using DCT transform

Threshold	*MSE*	*SNR*	*CR* [%]	NO
22.4	3.322	33.4176	92.365	720727

NO – Number of Operations

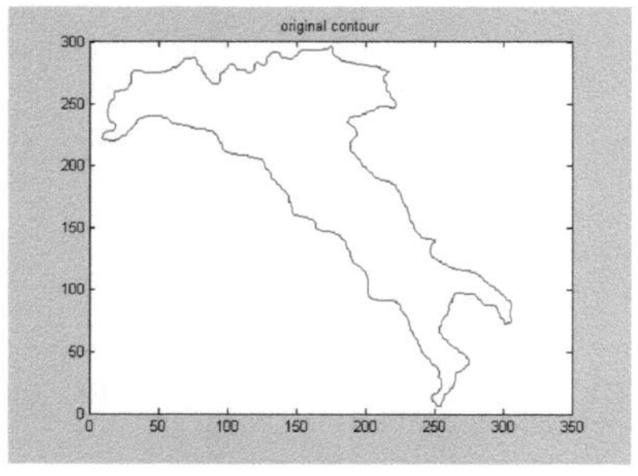

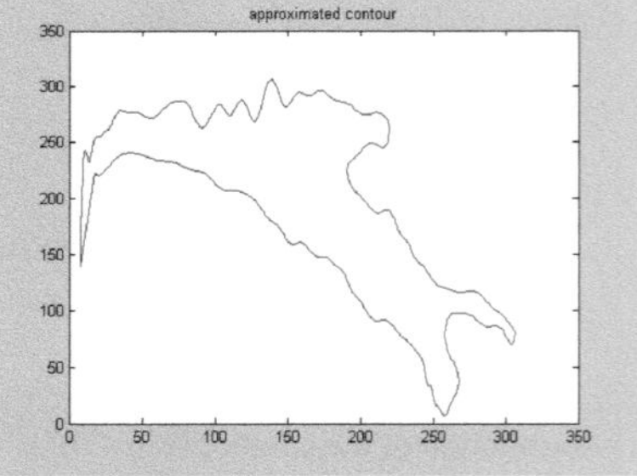

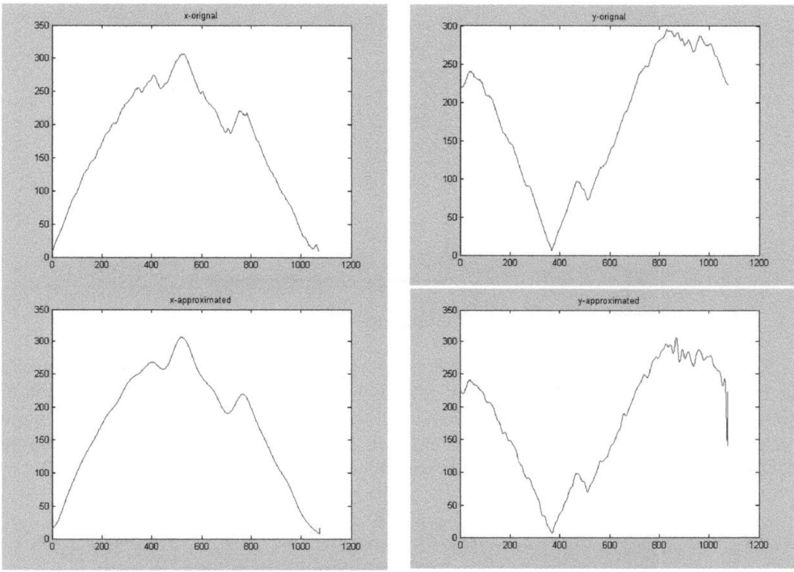

Fig. 4.10. Compressed contour using DCT transform.

4.4 Contour compression using Zonal Sampling

A lot of zonal sampling methods which was described in [51], shows that the best scheme for compression and contour extraction is as illustrated in Fig. 4.11. Fit criterion of the algorithm consists in selecting one of the squared block of the spectral images (e.g. shadow region) as LPF filter for image compression and the other coefficients will be taken into account in the contour reconstruction stage as shown in the Fig. 2. This algorithm in this work is referred as algorithm I [51], [52].

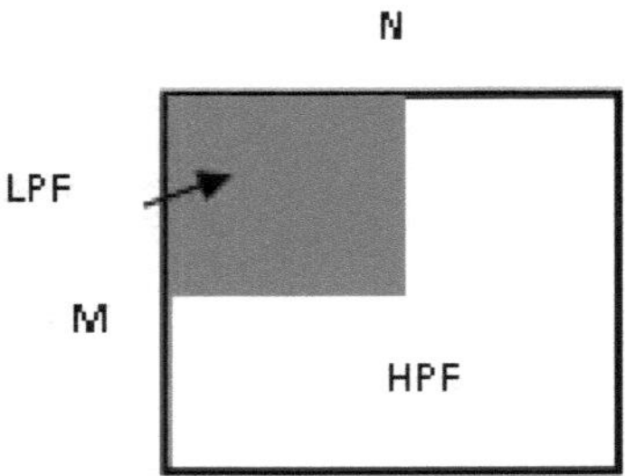

Fig. 4.11 Zonal sampling method using algorithm I

This paper compared this zonal method with a another zonal sampling method consists in selecting one block of the spectral images (i.e. shadow region) as LPF for image compression and the other coefficients will be taken into account in the contour reconstruction stage. This algorithm is referred as algorithm II and is shown in Figure 4.12 [52]. The block diagram of contour compression using DWT is shown in Fig. 4.13.

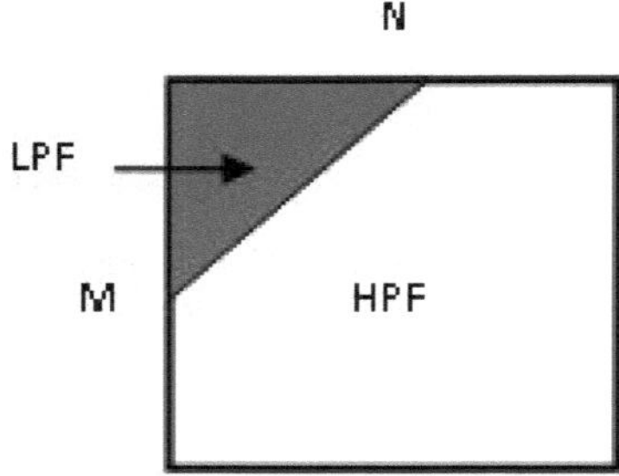

Fig. 4.12 Zonal sampling method using algorithm II

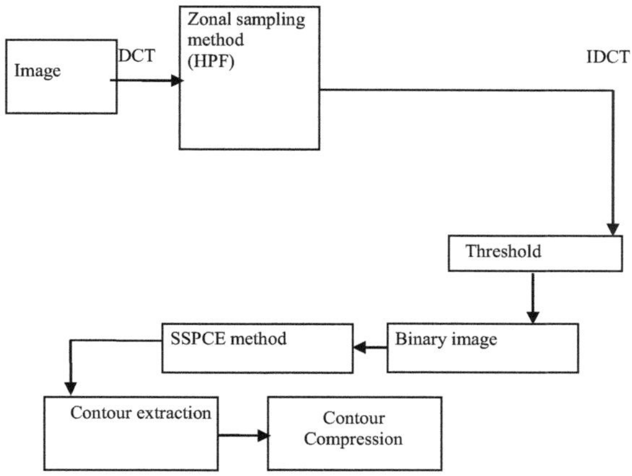

Fig. 4.13 Block diagram of the contour extraction & compression using DWT

>> A= imread ('tools.bmp'); >> figure;imshow(A,[]) % read and display the image ('tools') % see Fig. 4.14

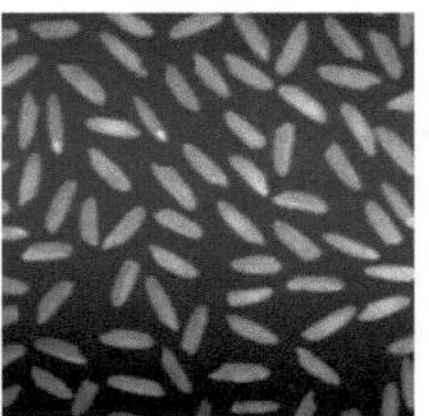

Fig. 4.14 'Tools' image

```
>> [n m]=size(A); %determine the image size
>> FT=dct2(A); % perform the dct
>> B = zeros (n,m); % create zeros matrix with same size of the image
>> B (11:n,11:m)= FT (11:n,11:m); % took the spectral image except the square 10x10
>> IT = idct2 (B); % perform the inverse dct
>> figure; imshow (IT,[]) % display the reconstruction image (see Fig. 4.15)
```

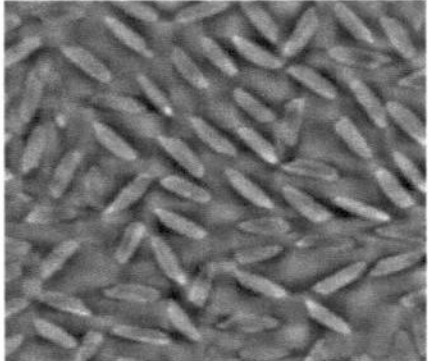

Fig. 4.15 The reconstructed image using High-pass filter

```
>> BW = im2bw (IT); % convert the reconstruction image to binary image
>> figure; imshow (BW,[]) % display the binary image (see Fig. 4.16)
```

Fig. 4.16 Binary image from HPF

```
>> C = SSPCE (BW); % contour extraction using SSPCE method
>> quit = CESSPCE (A,BW,C,n,m); % collecting the extracted contours as image
>> figure; imshow (quit,[]) % display the extracted contours as image (see Fig. 4.17)
```

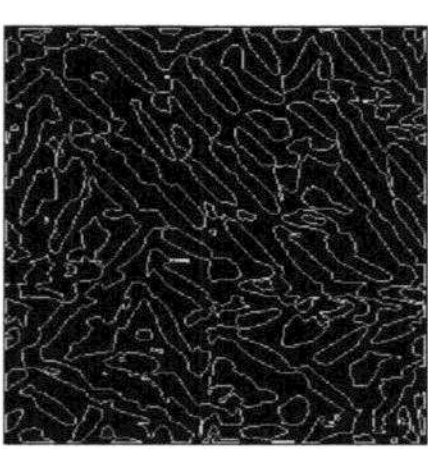

Fig. 4.17 Contour reconstruction using HPF

4.5 Haar Transform (HT)

The Haar transform uses rectangular basis functions and is the simplest of the wavelet transforms.

The basis functions are generated from the relations:

$$\mathrm{HA}(0,0,x) = \frac{1}{\sqrt{N}}$$

$$\mathrm{HA}(r,m,x) = \begin{cases} \dfrac{2^{\frac{r}{2}}}{\sqrt{N}} & \dfrac{m-1}{2^r} \le x < \dfrac{m-\frac{1}{2}}{2^r} \\ -\dfrac{2^{\frac{r}{2}}}{\sqrt{N}} & \dfrac{m-\frac{1}{2}}{2^r} \le x < \dfrac{m}{2^r} \\ 0 & elsewhere \end{cases} \tag{4.3}$$

over the range $0 \le x < 1$, where $0 \le r < \log_2 N$ and $1 \le m \prec 2^r$.

In the Haar transform most of the coefficients are functions of only part of the signal, e.g. the (orthonormal) 8 x 8 Haar transform matrix:

$$\frac{1}{\sqrt{8}}\begin{bmatrix} 1 & 1 & 1 & 1 & 1 & 1 & 1 & 1 \\ 1 & 1 & 1 & 1 & -1 & -1 & -1 & -1 \\ \sqrt{2} & \sqrt{2} & -\sqrt{2} & -\sqrt{2} & 0 & 0 & 0 & 0 \\ 0 & 0 & 0 & 0 & \sqrt{2} & \sqrt{2} & -\sqrt{2} & -\sqrt{2} \\ 2 & -2 & 0 & 0 & 0 & 0 & 0 & 0 \\ 0 & 0 & 2 & -2 & 0 & 0 & 0 & 0 \\ 0 & 0 & 0 & 0 & 2 & -2 & 0 & 0 \\ 0 & 0 & 0 & 0 & 0 & 0 & 2 & -2 \end{bmatrix}$$

Apart from powers of $\sqrt{2}$, all elements are 1, -1, and 0. One can interpret the multiplication of the Haar matrix with a signal vector as sampling the signal from low to

high frequencies (``sequences"). The first sample corresponds to the mean, the second to the mean difference between the first and last four neighbor pixels, and the last four transform samples represent differences of two neighboring pixels. Most of the transform coefficients depend only on their direct neighbors; the Haar transform thus has good high- and low-frequency response.
The Haar functions are shown in Fig. 4.18.

The Haar Transform equations is as follows

$$[S(n)] = \frac{1}{N}[Har(n)].[X(n)] \quad for \quad n = \log_2 N \tag{4.4}$$

$$[X(n)] = [Har(n)]^{-1}.[S(n)] \tag{4.5}$$

where: $[S(n)]$ - vector of spectral components and $[Har(n)]$ - Haar matrix

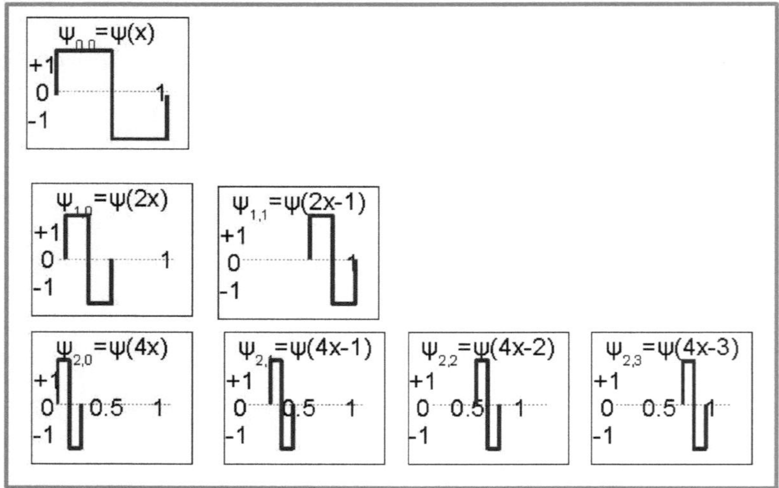

Fig. 4.18 Set of Haar functions

Fig. 4.19 shows that the Haar transform is performed for each of X(n) and Y(n); then the threshold sampling is done and finally the inverse Haar is applied for each X and Y of Haar to reconstruct the approximated contour (the related results are shown in Table 4.2).

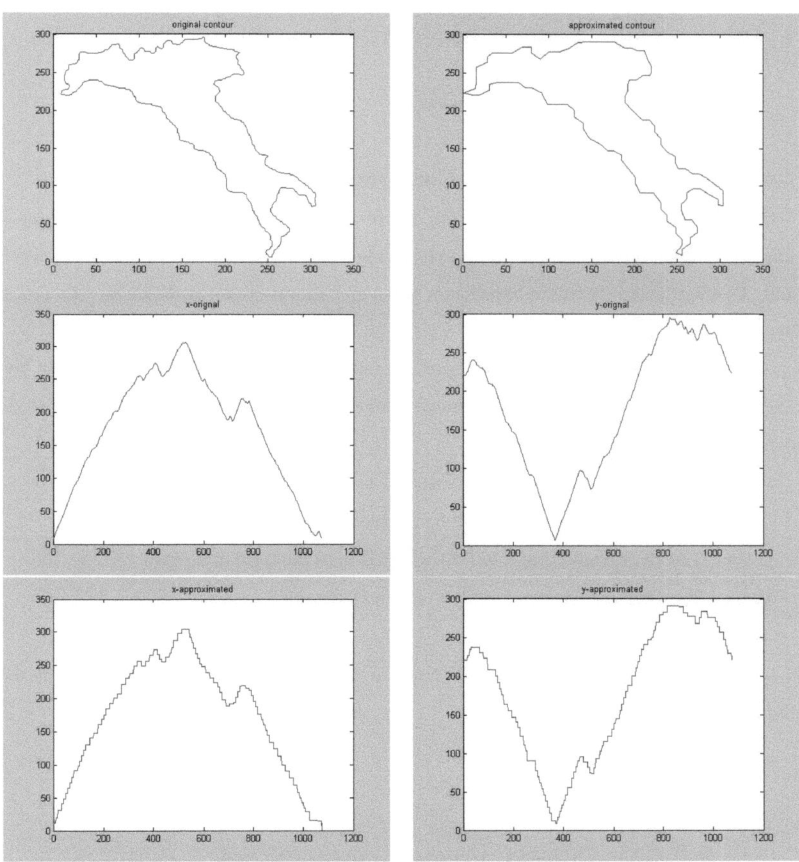

Fig. 4.19 Compressed contour using HT

Table 4.2 Results of approximation for the Italy contour using HT transform

Threshold	*MSE*	*SNR*	*CR* [%]	NO
0.325	3.3249	33.4137	91.4804	99151

NO – Number of Operations

4.6 Time & Spectral Domains Compression Methods Comparison

The computational complexity is one of the most important factors in evaluating a transform. High computational complexity leads to high implementation cost. The MSE and SNR criterions versus compression ratio are also used to evaluate the distortion between the purpose comparisons.

The comparison is done for Italy contour which was extracted by the using of the "SSPCE" (Single Step Parallel Contour Extraction) [11]. The comparison is made between the following algorithms:

- Ramer method; it will be referred to as the first algorithm.
- The Triangle method; it will be referred to as the second algorithm.
- The DCT transform; it will be referred to as the third algorithm.
- The Haar transform; it will be referred to as the fourth algorithm.

The Comparison of the compression abilities versus the MSE & SNR are shown in the Fig. 4.20 & Fig. 4.21 respectively.

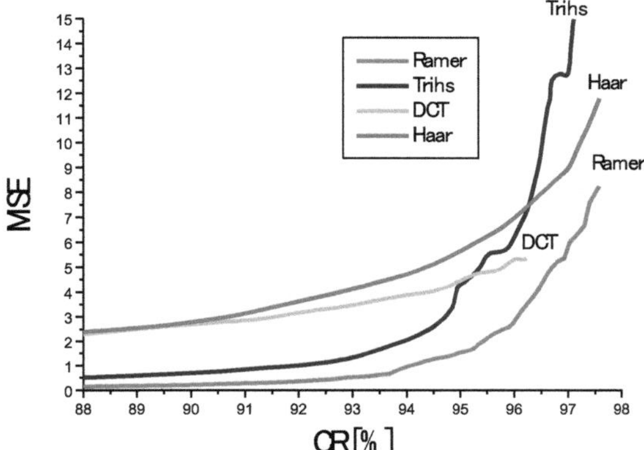

Fig. 4.20 MSE versus CR for the Italy contour

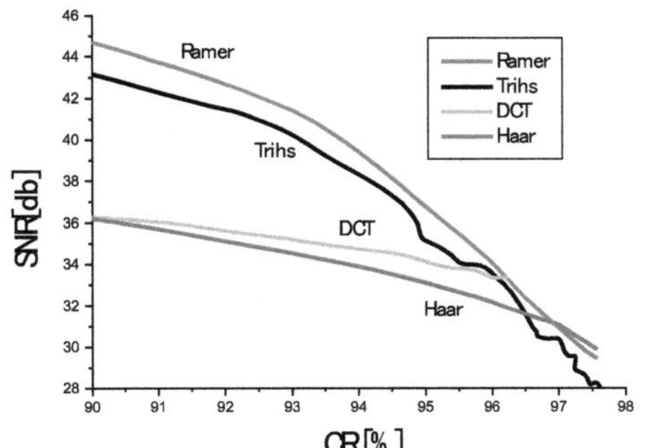

Fig. 4.21 SNR versus CR for the Italy contour

From the Fig. 4.20 & Fig 4.21; Trihs algorithms are faster than that of Ramer method with high compression ratio.

The DCT is the slowest one. Trihs is very close to the Ramer method for high compression ratio. The reconstruction quality by DCT is not so good at the higher compression ratio.

4.7 Discrete Wavelet Transform (DWT)

The Wavelet analysis is an exciting new method for solving difficult problems in mathematics, physics, and engineering, with modern applications as diverse as wave propagation, data compression, signal processing, image processing, pattern recognition, computer graphics, the detection of aircraft and submarines and other medical image technology [59] [60]. Wavelets allow complex information such as music, speech, images and patterns to be decomposed into elementary forms at different positions and scales and subsequently reconstructed with high precision.
Wavelets are obtained from a single prototype wavelet called mother wavelet by dilations and shifting using the equation (1).

$$\Psi_{a,b}(t) = \frac{1}{\sqrt{a}} \Psi(\frac{t-b}{a}) \tag{4.6}$$

This paper presents a method of contour extraction and compression from grey level image. The proposed algorithm is applied in spectral domain using single-level of discrete wavelet transform (DWT). Single step parallel contour extraction (SSPCE) method is used for the binary image after inverse wavelet transform is applied to the details images. Then the contours are compressed using either Ramer, Tangent, Triangle, or Trapezoid methods in spatial domain. Effectiveness of the contour extraction and compression for different classes of images is evaluated. In the paper the main idea of the analyzed procedure for both contour extraction and image compression are performed. To compare the results, the mean square error, signal-to-noise ratio criterions, and compression ratio (or bit per pixel) were used. The simplicity to obtain compressed image and extracted contours with accepted level of the reconstruction is the main advantage of the proposed algorithm.

Contour representation and compression are required in many applications e.g. computer vision, topographic or weather maps preparation, medical images and moreover in image compression. The transform coding method compresses image data by representing the original signal with a small number of transform coefficients. It exploits the fact that for typical images a large amount of signal energy is concentrated in a small number of coefficients. The goal of transform coding is to minimize the number of retained transform coefficients while keeping distortion at an acceptable level. Transform coding is an integral part of one of the most widely known standards for lossy image compression, the JPEG (Joint Photographic Experts Group) standard. Contour extraction and image compression can be obtained using transforms such as Fourier [53], Walsh

[51], DCT [55], Wavelet [59], [60] and periodic haar piecewise-linear (PHL) which is based on the integration of haar functions [56], [57]. In this paper the discrete wavelet transform will be used. To obtain the compressed image and binary image, inverse wavelet transform is applied to the approximation coefficients image and details coefficients images respectively. The contours are extracted from binary image using single step parallel contour extraction (SSPCE) method [54], [58]. Finally the compressed contours are obtained using either Ramer, Tangent, Triangle, or Trapezoid methods.

Flowchart of the algorithm for image compression and contour extraction is depicted in Fig. 4.22.

4.8 Applied Measures

The proposed image compression and contour extraction method is related to the data compression and extraction problems. To evaluate its compression ability, the following compression ratio (or bit per pixel) was introduced if each pixel is implemented by eight bits, and is defined by equations (4.7) and (4.8) respectively.

$$CR = \frac{NOZ}{(n*m)} * 100\% \tag{4.7}$$

where: NOZ is number of zero coefficients in the spectral domain; $n * m$ is size of the image

$$bpp = \frac{ZS * 8}{(n*m)} \tag{4.8}$$

where: ZS is coefficients number in the desired zonal

The mean square error (MSE) and peak signal-to-noise ratio (PSNR) criterions were used to evaluate the distortion introduced during the image compression and contour extraction procedures, and is defined by equations (4.9) and (4.10) respectively.

$$MSE(I,\tilde{I}) = \frac{1}{(n*m)} \sum_{i=0}^{n} \sum_{j=0}^{m} (I(i,j) - (\tilde{I}(i,j))^2 \tag{4.9}$$

$$PSNR(I,\tilde{I}) = 10\log_{10} \frac{(L-1)^2}{MSE(I,\tilde{I})} \tag{4.10}$$

where I is the original image, $\tilde{I}$ is the reconstructed image, and L is the grey-level number.

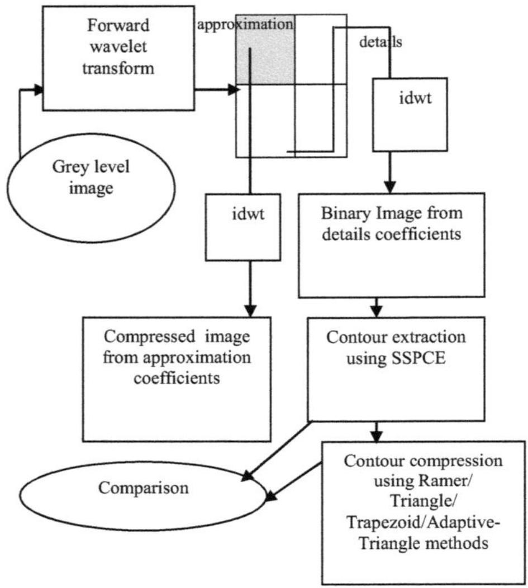

Fig. 4.22. Flowchart of the proposed algorithm for image compression and contour compression

4.9 Results of the Experiments

To visualize the experimental results a 'Tool' grey levels images were selected. Selected image are shown in Fig. 4.23.

>>A=imread('tools.bmp'); % read the 'Tools' image
[X,map]=gray2ind(A, N); % converts the intensity image A to an indexed image X with colormap GRAY(N). If N is omitted, it defaults to 64.
>>[X,map]=gray2ind(A);
>>X=double(X);[n,m]=size(A); % store X in double format and determine its size

>> figure; imshow (X, map) % display X image with its colour map (see Fig. 4.23)

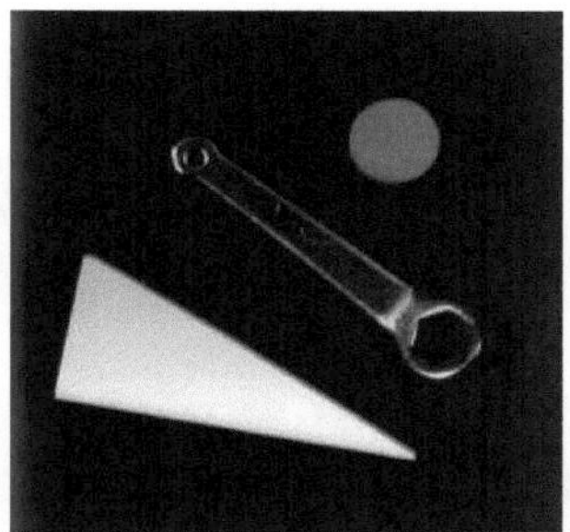

Fig. 4.23. Test images: Tools (256x256)

The decomposition of Tools image using first level of DWT is shown in Fig. 4.24.

```
[wc,s] = wavedec2(A,level,wname); % Multilevel 2-D wavelet decomposition
```

This command returns the wavelet decomposition of the matrix A at level N, using the wavelet named in string 'wname' (see WFILTERS) like Haar, Daubechies, Coiflets and etc. Outputs are the decomposition vector wc and the corresponding bookkeeping matrix S. N must be a strictly positive integer.

If we use 'Haar' wavelet with single (first) level then we should write

```
>>[wc,s] = wavedec2(A,1,'haar');
% Extract the level 1 coefficients.
a1 = appcoef2(wc,s,wname,1); % approximation image
h1 = detcoef2('h',wc,s,1);     % horizontal image
v1 = detcoef2('v',wc,s,1);      % vertical image
d1 = detcoef2('d',wc,s,1);      % details image

% Display the decomposition up to level 1 only.
ncolors = size(map,1);          % Number of colors.
sz = size(X);
cod_a1 = wcodemat(a1,ncolors); cod_a1 = wkeep(cod_a1, sz/2);
cod_h1 = wcodemat(h1,ncolors); cod_h1 = wkeep(cod_h1, sz/2);
cod_v1 = wcodemat(v1,ncolors); cod_v1 = wkeep(cod_v1, sz/2);
cod_d1 = wcodemat(d1,ncolors); cod_d1 = wkeep(cod_d1, sz/2);
```

```
>> figure; imshow([cod_a1,cod_h1;cod_v1,cod_d1],[]); % display the first level of forward wavelet transform (Fig. 4.24)
```

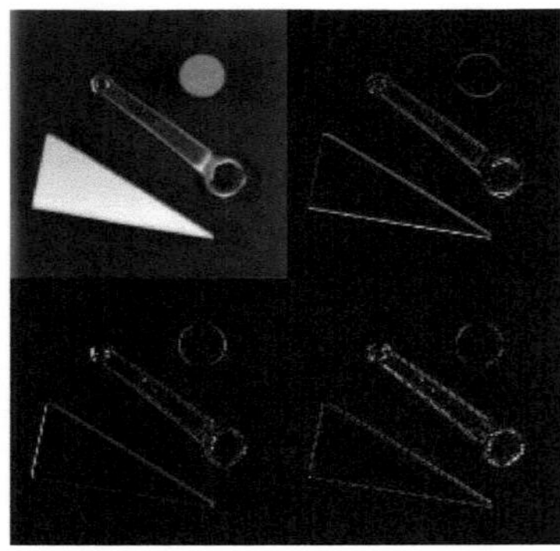

Fig. 4.24. Tools image decomposition using first level of DWT

```
>> ra1 = wrcoef2('a',wc,s,'haar',1);

>> rh1 = wrcoef2('h',wc,s,'haar',1);

>> rv1 = wrcoef2('v',wc,s,'haar',1);

>> rd1 = wrcoef2('d',wc,s,'haar',1);
```

```
>>cod_ra1 = wcodemat(ra1,ncolors);

>>cod_rh1 = wcodemat(rh1,ncolors);

>>cod_rv1 = wcodemat(rv1,ncolors);

>>cod_rd1 = wcodemat(rd1,ncolors);
```

```
>>figure; % reconstructed sub images of first level
```

```
>>subplot(3,4,1); imshow(X,[]); axis image; set(gca,'XTick',[],'YTick',[]); title('Original')

>>subplot(3,4,5); imshow(cod_ra1,[]); axis image; set(gca,'XTick',[],'YTick',[]); title('ra1')

>>subplot(3,4,6); imshow(cod_rh1,[]); axis image; set(gca,'XTick',[],'YTick',[]); title('rh1')

>>subplot(3,4,7); imshow(cod_rv1,[]); axis image; set(gca,'XTick',[],'YTick',[]); title('rv1')

>>subplot(3,4,8); imshow(cod_rd1,[]); axis image; set(gca,'XTick',[],'YTick',[]); title('rd1')

% see Fig. 4.25
```

The binary image can be obtained using thresholding as follows (see Fig.4.26):

```
>> BWrh1=cod_rh1>4; % binary image from horizontal image (threshold value here is 4)

>> BWrv1=cod_rv1>4; % binary image from vertical image (threshold value here is 4)

>> BWrd1=cod_rd1>6; % binary image from details image (threshold value here is 6)

% Display the binary image from horizontal, vertical and details images

>> subplot(1,3,1); imshow(BWrh1,[]); axis image; set(gca,'XTick',[],'YTick',[]); title('Horizontal Coefficients')

>> subplot(1,3,2); imshow(BWrv1,[]); axis image; set(gca,'XTick',[],'YTick',[]); title('Vertical Coefficients')
```

Original

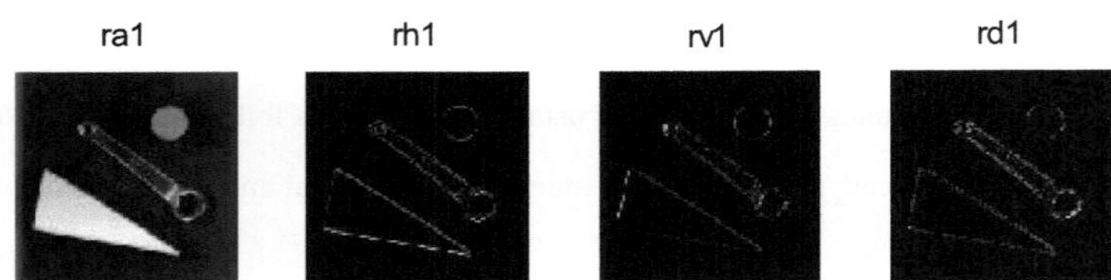

Fig. 4.25. First level sub images using DWT

>> subplot(1,3,3); imshow(BWrd1,[]); axis image; set(gca,'XTick',[],'YTick',[]); title('Details Coefficients')

% The contours of these binary images can be extracted using SSPCE method using the commands

>> Ch=SSPCE(BWrh1); % contours extracted from horizontal binary image

>> CE_h = CESSPCE(A,BWrh1,Ch,256,256); % collected the horizontal extracted contours image where 256, 256 means the image size

>> figure; imshow(CE_h,[]) % display the extracted contours image of horizontal image (see Fig. 4.27

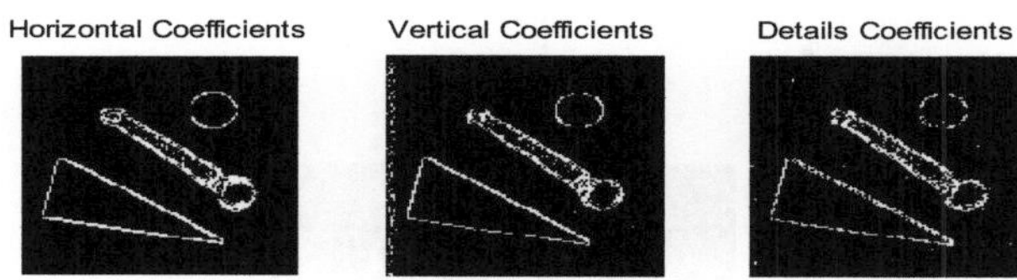

Fig. 4.26. Binary image using thresholding values & first level of DWT

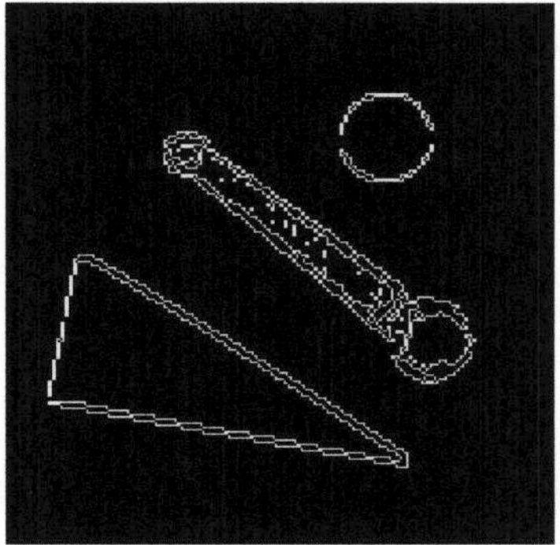

Fig. 4.27. Contour extraction using Horizontal image of first level of DWT

>> Cv=SSPCE(BWrv1); % contours extracted from vertical binary image

>> CE_v = CESSPCE(A,BWrv1,Cv,256,256); % collected the vertical extracted contours image where 256, 256 means the image size

>> figure; imshow(CE_v,[]) % display the extracted contours image of vertical image (see Fig. 4.28)

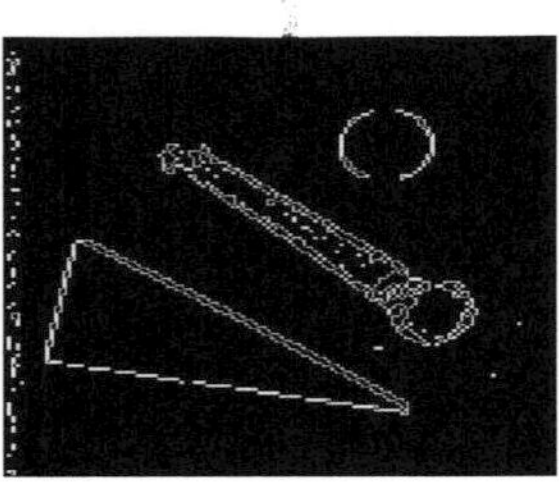

Fig. 4.28. Contour extraction using Vertical image of first level of DWT

>> Cd=SSPCE(BWrd1); % contours extracted from diagonal binary image

>> CE_d = CESSPCE(A,BWrd1,Cd,256,256); % collected the extracted contours image where 256, 256 means the image size

>> figure; imshow(CE_d,[]) % display the extracted contours image of diagonal image (see Fig. 4.29)

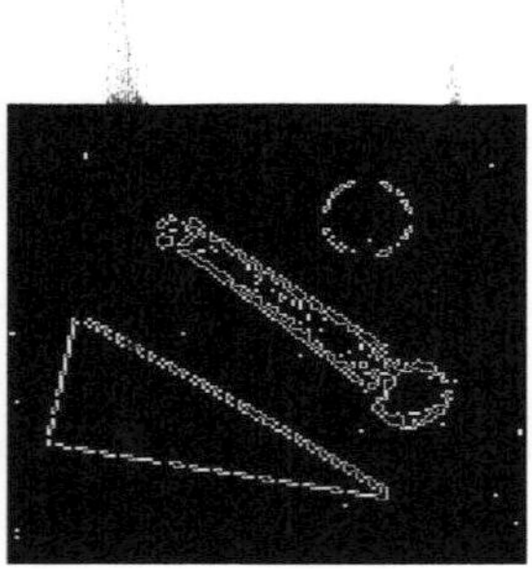

Fig. 4.29. Contour extraction using Diagonal image of first level of DWT

Adding together the reconstructed average at level 1 and all of the reconstructed details gives the full reconstructed image.

>> Xhat = ra1 + rh1 + rv1 + rd1;

>> figure; imshow(Xhat,[]) % display the reconstructed image (see Fig. 4.30)

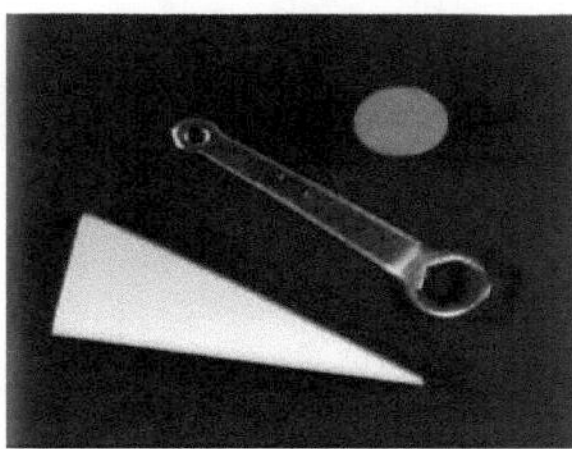

Fig. 4.30. Reconstructed image from level 1 (approximation & all of the reconstructed details)

Another way to reconstruct the image using the command

>> XXhat = waverec2(wc,s,wname);

We can check wither both different way of the reconstructed image are equals you may write

>> isequal (Xhat, XXhat) % if they are equals the answer will 1 otherwise o

ans =

1

The extracted contours using SSPCE method for contour extraction of Tools image are obtained using horizontal, vertical, and diagonal coefficients as shown in Fig. 4.31.

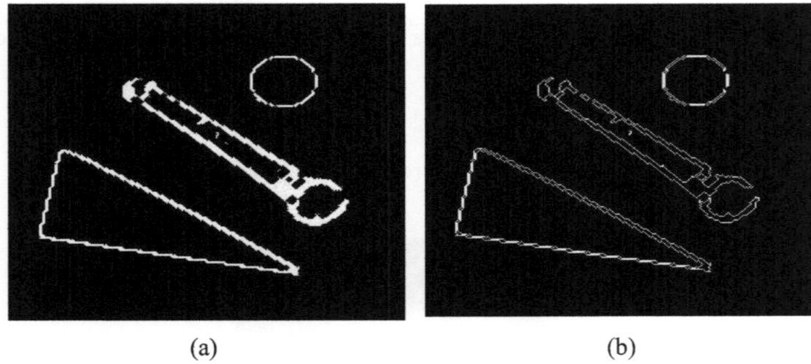

(a) (b)

Fig. 4.31. Tools image: (a) Binary image from details coefficients, and (b) Contours extraction using SSPCE method

The compressed contours for Tools image are obtained using Ramer, Tangent, Triangle, and Trapezoid methods are shown in Fig. 4.32 (related results are shown in the Table 4.3).

The proposed algorithm is compared with binary image which is obtained using suitable threshold criteria as shown in Fig. 4.33 (related results are shown in Fig. 4.34 and in Table 4.4).

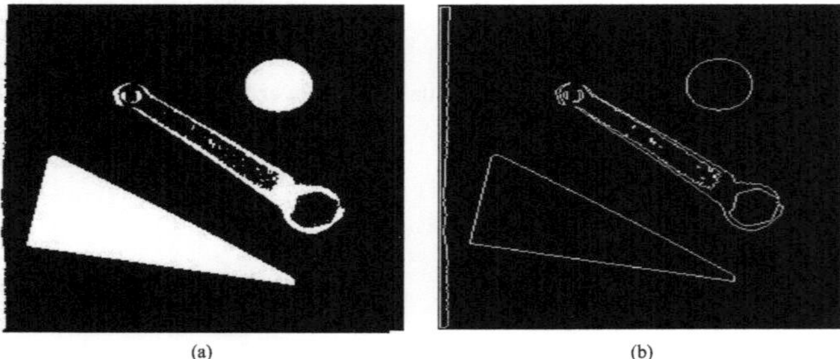

(a) (b)

Fig. 4.33. Tools image: (a) Binary image using threshold, and (b) Contours extraction using SSPCE method

Table 4.3

Measures **Method (Compression)**	*MSE*	*PSNR [db]*	*CR*	*Elapsed Time*
a) Tangent	0.0175	17.58	69.46	9.17
b) Ramer	0.0175	17.58	69.46	10.65
c) Triangle	0.0172	17.65	68.43	9.13
d) Trapezoid	0.0174	17.59	69.34	10.40
e) Tangent	0.0199	17.01	79.24	7.98
f) Ramer	0.0199	17.02	78.99	8.82
g) Triangle	0.0202	16.95	80.39	7.39
h) Trapezoid	0.0197	17.05	78.57	8.31
i) Tangent	0.0226	16.47	89.80	6.63
j) Ramer	0.0226	16.47	89.80	7.35
k) Triangle	0.0226	16.47	89.80	6.48
l) Trapezoid	0.0225	16.47	89.62	7.27

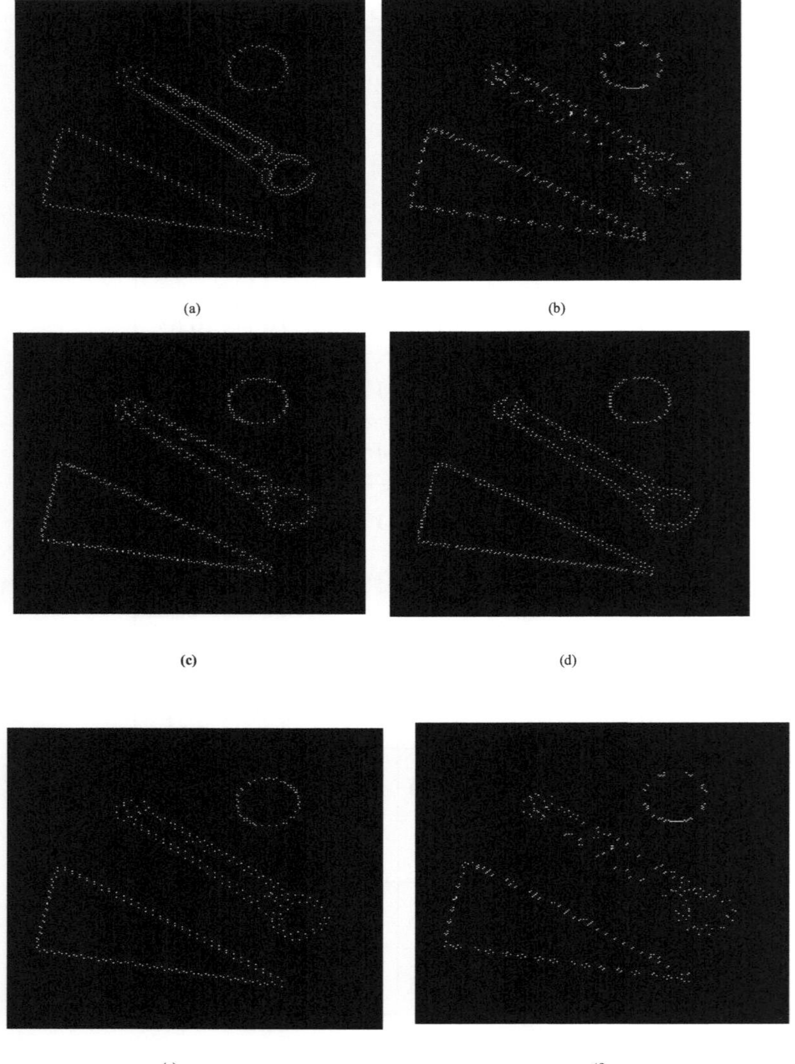

(a) (b) (c) (d) (e) (f)

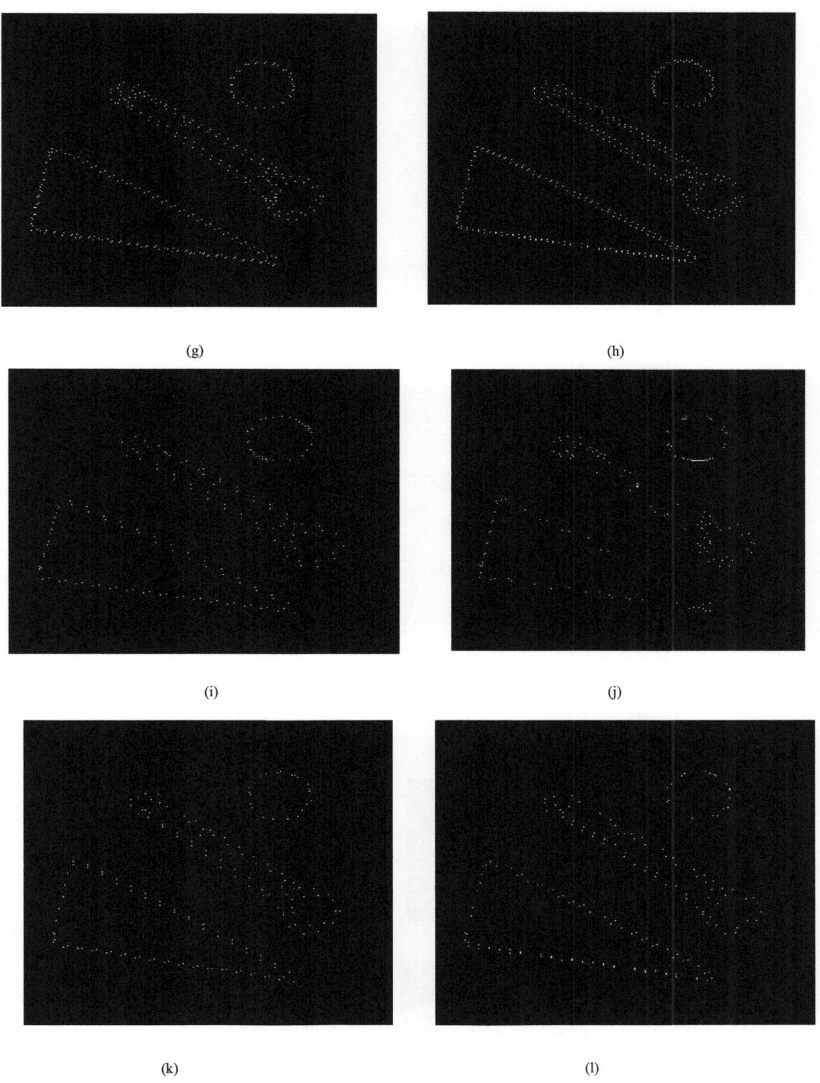

(g) (h)

(i) (j)

(k) (l)

Fig. 4.32 Tools image contour compression using Ramer, Tangent, Triangle, and Trapezoid methods

Table 4.4

Measures **Method (Compression)**	*MSE*	*PSNR [db]*	*CR*	*Elapsed Time*
a) Tangent	0.0176	17.54	67.09	10.45
b) Ramerb	0.0183	17.38	69.47	9.89
c) Triangle	0.0184	17.36	69.88	9.44
d) Trapezoid	0.0182	17.39	69.36	9.13
e) Tangent	0.0206	16.86	78.29	8.52
f) Ramer	0.0207	16.84	78.82	8.59
g) Triangle	0.0205	16.88	78.06	8.36
h) Trapezoid	0.0205	16.88	78.06	8.09
i) Tangent	0.0235	16.28	89.50	7.11
j) Ramer	0.0236	16.27	89.84	7.04
k) Triangle	0.0235	16.28	89.55	6.91
l) Trapezoid	0.0236	16.28	89.67	6.58

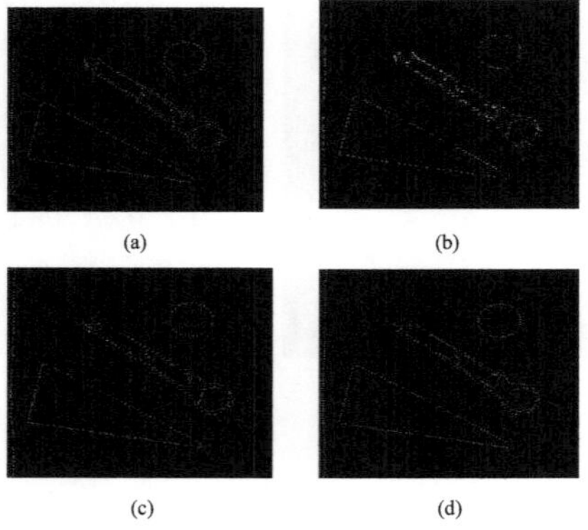

(a) (b)

(c) (d)

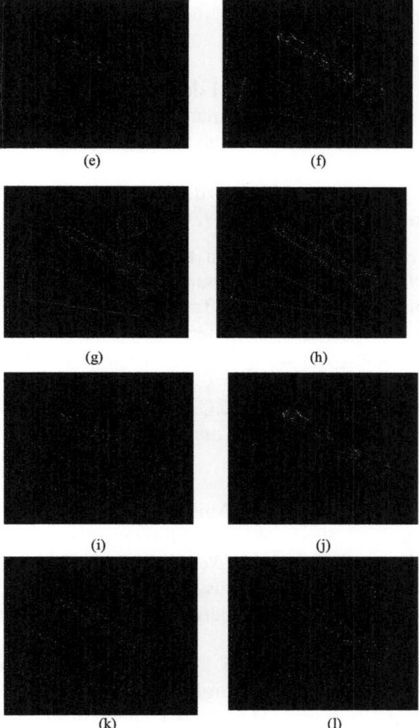

Fig. 4.34 Tools image contour compression using Ramer, Tangent, Triangle, and Trapezoid methods (by threshold)

The good quality of contour extraction and compression are the main advantage of the proposed algorithm compared with the binary image using suitable threshold value. By using single level of discrete wavelet transform the two sub-images are obtained (compressed image and extracted contour). Ramer, tangent, triangle, and trapezoid methods are used to compress the extracted contours without significant visible distortion. The reconstruction quality improvement of compressed contour compared to the binary image which is obtained by using of suitable threshold value is about 0.2 decibels. The contour compression using triangle and tangent methods is much faster than using of Ramer and Trapezoid method for all analyzed cases. Important advantage of the analyzed method is the simplicity of implementation both in terms of memory requirement and fit criterion complication.

References

[1] Gonzalez, R. C., Woods, R. E., & Eddins, S. L. (2009). Digital image processing using MATLAB (2nd Ed.). Gatesmark Publishing.

[2] Marques, O. (2011). Practical image and video processing using MATLAB. Wiley.

[3] L.S. Davis, "A survey of edge detection techniques", *Computer Vision, Graphics, and image processing 4 (1975),* pp. 248-270.

[4] R.M. Haralick, "Edge and region analysis for digital image data", *Computer Vision, Graphics and Image Processing (1980),* pp. 60-73.

[5] Sadayuki Hongo, Mitsuo Kawato, Toshi Inui and Sei Miyake, "Contour extraction of images on parallel computer-local, Parallel and stochastic algorithm which learns energy parameters", *International Joint Conference on Neural Networks (Washington D.C.), 1989,* pp. 161-168.

[6] P.K. Sahoo, S. Soltani, A. K.C. Wong and Y.C. Chen, "A survey of thresholding techniques", *Computer Vision, Graphics and Image Processing (1988),* pp. 233-260.

[7] Marie-Pierre Dubuisson and Anil K. Jain, "Object contour extraction using color and motion", *Proc. IEEE Computer Society Conference on Computer Vision and Pattern Recognition (New York), 1993,* pp. 471-476.

[8] Safar, Maytham H., Shahabi, Cyrus, "Shape Analysis and Retrieval of Multimedia Objects", Springer Berlin Heidelberg New York, 2003, Hardcover ISBN: 1-4020-7252-X.

[9] A. Nabout and H. A. Nour Eldin, "The topological contour closure requirement for object extraction from 2D-digital images", Proc. 1993 IEEE International Conference on Systems, Man and Cybernetics (Le Touquet, France), 1993, pp. 120-125.

[10] Nabout A., Su B., Nour Eldin H. A., "A Novel Closed Contour Extractor, Principle and algorithm", Proc. Of the International IEEE/ISCAS Conf. on Circuits and Systems, April 29 – May 3, Seattle, USA, Vol.1, pp. 445-448, 1995.

[11] A. Dziech, W. S. Besbas, "Fast Algorithm for Closed Contour Extraction", proc. of the Int. Workshop on Systems, Signals and Image Processing, Poznań, Poland, 1997, pp. 203-206.

[12] W. Besbas, "Contour Extraction, Processing and Recognition", Poznan University of Technology, Ph. D. Thesis, 1998.

[13] Didier Demigny and Tawfik Kamle, "A discrete expression of cranny's criteria for step edge detector performances evaluation", Proc. IEEE Transactions on Pattern Analysis and Machine Intelligence 19 (1997), pp. 1199-1211.

[14] Michael Leyton , "A Generative Theory of Shape", Springer-Verlag, October 2001.

[15] M. Bennamoun, B. Boashash, and A. Bower, "Object contour extraction from motion parameters", Proc. 2nd Australian and New Zealand Conference on intelligent Information Systems (Brisbane, Australia), pp. 337-341.

[16] M. Fleck, "Some defects in finite-difference edge finders", *Proc. IEEE*

Transactions on Pattern Analysis and Machine Intelligence 14 (1992), pp. 337-345.

[17] P. Parodi and G. Piccioli, “A feature-based recognition scheme for traffic scenes”, *Proc. Of the Intelligent Vehicles (Detroit, USA), 1995.*

[18] Gerhard X. Ritter and Joseph N. Wilson, “Computer vision algorithms in image algebra”, CRC Press, New York, 1996.

[19] Giovanni Marola and Anna Vaccarelli, “An algebra method for detecting three-dimensional objects in a picture”, *Proc. IAPR International Conference on Pattern Recognition (The Hague, The Netherlands), 1992,* pp. 415-419.

[20] Peter J. Olver, Guillermo Sapiro and Allen Tannenbaum, “A ffine Invariant detection: Edges, active contours and segments”, *Proc. IEEE Computer Society Conference on Computer Vision and Pattern Recognition (San Francisco, California), 1996,* pp. 520-525.

[21] R. Mohrotra, “Similar-shape retrieval in shape data management”, *Computer 28 (1995),* pp. 57-62.

[22] Wei Xiong and Christine Grafffigne, “A hierarchical method for detection of moving objects”, Proc. IEEE International Conference on Image Processing (Austin, Texas), 1994, pp. 795-799.

[23] JL. Starck and F. Murtagh, “Astronomical Image and Data Analysis”, Springer, 2006 (2nd edition).

[24] Krim, Hamid; Yexxi, Jr., Anthony, “Statistics and Analysis of Shapes’, 2006, XL 395 p., 143 illus., Hardcover ISBN: 0-8176-4376-1.

[25] Batchelor B. G., Laing S. G., “Polar-Vector Representations of edges in Pictures” *Electronics Letters, Vol.13, No.24,* pp.727-729, 1977.

[26] Freeman H., “Boundary Encoding and processing”, *in Picture Processing and Psychopictorics (B. S. Lipkin and Rosenfeld, Eds),* pp.241-266, Academic Press, NY, 1970.

[27] Scott E. Umbaugh, “Computer vision and image processing”, Prentice-Hall, 1998.

[28] Nalini K. Ratha, Tolga Acar, Muhittin Gokmen, and Anil K. Jain, “A distributed Edge detection and surface reconstruction algorithm”, *Proc. Computer Architectures for machine Perception (Como, Italy), 1995,* pp. 149-154.

[29] Yali Amit, “2D Object Detection and Recognition”, MIT Press, 2002.

[30] G. Economou, S. Fotopoulos, and M. Vemis, “A noval edge detector based on non – linear local operations”, *Proc. IEEE International Symposium on Circuits and Systems (London), 1994,* pp. 293-296.

[31] D.H. Ballard and C.M. Brown, “Computer vision”, Prentice Hall, Englewood Cliffs, NJ, 1982.

[32] Duda R.O., and Hart P.E., “Pattern Classification and Scene Analysis”, John Wiley & Sons, New York, 1973.

[33] Heijden F., “Image Based Measurement Systems”, John Wiley, 1995.

[34] U. Ramer, “An Iterative Procedure for the Polygonal Approximation of Plane Curves”, Computer Graphics and Image Proc., Academic Press, 1972, pp.244-256.

[35] A. Dziech, R. Baran, “The Tangent Method Algorithms for Contour Approximation”, *International Workshop on Multimedia Communication*

and Services, Cracov, Poland, 2004, pp. 11-16.

[36] R. Baran, W. Dziech, "Tangent Method of Contour Approximation", *International Workshop on Multimedia Communication and Services*, Kielce, Poland, 2003, pp. 11-16.

[37] Andrzej Dziech, Ali Ukasha and Remigiusz Baran, "A New Method For Contour Compression", *WSEAS Int. Conf. on Signal, Speech and Signal Processing (SSIP 2005),* Corfu Island, Greece, August 17-19, 2005, pp.282-286.

[38] Andrzej Dziech, Ali Ukasha and Remigiusz Baran, "Fast Method For Contour Approximation And Compression", *WSEAS Transactions on Communications Issue 1, Volume 5,* January 2006, pp. 49 - 56.

[39] A. Ukasha, A. Dziech, E. Elsherif and R. Baran, An efficient method of contour compression, *International Conference on Visualization, Imaging and Image Processing (IASTED/VIIP),* Cambridge, United Kingdom, 2009, pp. 213-218.

[40] A. Ukasha, A New Method for Contour Approximation using Basic Ramer Idea, *International Conference on Pattern Recognition and Computer (WASET/*ICPRC) Dubai, United Arab Emirates, 2011.

[41] A. Ukasha, Arabic Letters Compression using New Algorithm of Trapezoid method, *International Conference on Signal Processing, Robotics and Automation (ISPRA),* 2010, pp. 336-341.

[42] Majdi F. M. EL Bireki, Ruqiya M. A. ALI, ALI A. M. Ukasha, Mohammed F. L. ABDULLAH, An Adaptive-Triangle Method for Binary Image Contour Data Approximation, International Journal of Theoretical and Applied Information Technology (JATIT), Volume 81, Issue 2, pp. 343-348, 2015. ISSN 1992-8645.

[43] Vetterli, Martin & Kovacevic, Jelena, Wavelets and Subband Coding, Printice Hall Inc., 1995.

[44] Walsh, J. L. "A Closed Set of Normal Orthogonal Functions", *Amer. J. Math.* 45, 5-24, 1923.

[45] Wolfram, S., "A New kind of Science", Champaign, IL: Wolfram Media, pp. 573 and 1072-1073, 2002.

[46] Harmuth, H. F., "Applications of Walsh Functions in Communications", *IEEE Spectrum* 6, pp.82-91, 1969.

[47] Gonzalez, R.C. and Woods, R. E., "Digital Image Processing", Addison – Wesley, Reading, MA., 1992.

[48] A. Dziech, F. Belgassem & H. J. Nern, Image data compression using zonal zonal sampling and piecewise-linear transforms, *Journal of Intelligent And Robotic Systems. Theory & Applications,* 28(1-2), Kluwer Academic Publishers, June 2000, 61-68.

[49] Brigham E.O., "The Fast Fourier Transform", Prentice-Hall, Englewood Cliffs, 1974.

[50] Clarke R. J., "Transform Coding of Images", Academic Press, 1985.

[51] Didier Demigny and Tawfik Kamle, "A discrete expression of cranny's criteria for step edge detector performances evaluation", *Proc. IEEE Transactions on Pattern Analysis and Machine Intelligence 19 (1997),* pp. 1199-1211.

[52] Hemant D. Tagare and Rui J.P. deFigueiredo, "Reply to on the localization

performance measure and optimal edge detection", *Proc. IEEE Transactions on Pattern Analysis and Machine Intelligence 16 (1994),* pp. 108-110.

[53] E. O. Brigham, *The Fast Fourier Transform*, Prentice-Hall, Englewood Cliffs, 1974.

[54] W. Besbas, Contour Extraction, Processing and Recognition, Poznan University of Technology, Ph. D. Thesis, 1998.

[55] R. J. Clarke, Transform Coding of Images, Academic Press, 1985.

[56] A. Dziech, F. Belgassem and H. J. Nern, Image data compression using zonal sampling and piecewise-linear transforms, Journal of Intelligent And Robotic Systems, Theory & Applications, 28(1-2), 2000, pp. 61-68.

[57] A. Dziech, W. Besbas, A. Nabout, and H. Nour Eldin, Fast algorithm for closed contour extraction, Proc. of the 4th International Workshop on Systems, Signals and Image Processing, Poznan, Poland, 1997, pp. 203-206.

[58] A. Ukasha, A. Dziech, E. Elsherif, and R. Baran, An efficient method of contour compression, International Conference on Visualization, Imaging and Image Processing (IASTED/VIIP), 2009, pp. 213-218.

[59] R. C. Gonzalez, Digital Image Processing, Second Edition, Addison Wesley, 1987.

[60] M. Vetterli, and J. Kovacevic, Wavelets and Subband Coding, Prentice-Hall, Englewood Cliffs, NJ, 1995.

[61] A. Dziech, A. Ukasha, and J. Wassermann, A New Method for Contour Extraction and Image Compression in Spectral Domain, International Conference on Multimedia, Signal processing and Communications, Zadar, Croatia,pp. 41 – 44 (2006).

[62] A. Ukasha, An Efficient Zonal Sampling Method for Contour Extraction and Image Compression using DCT Transform, The 3rd Conference on Multimedia Computing and Systems (ICMCS'12), Tangier, Morocco, May, 2012.

[63] R. C. Gonzalez and R. E. Woods, Digital Image Processing, Third Edition, Addison Wesley, 2008.

Appendixes

```
Function ContourCollection = SSPCE(image);
%
%SSPCE - Single Step Parallel Contour Extraction
%image  - binary input mage
% Author: Dr. Ali Ukasha
ContourCollection = {};
contour = 1;      tlo = 0;
k = contour;      t = tlo;
%------
%step 1 -  putting zeros to the border line
[n,m] = size(image);
b = zeros(n,m);
b(2:n-1,2:m-1) = image(2:n-1,2:m-1);
%------
%step 2 -  contour extraction
a = zeros(n,m); %the table of directional codes
for j = 2:n-1,
   for i = 2:m-1,
      if (b(j,i) == 1),
      while (a(j,i) == 0),
        if ( ((b(j,i+1) & b(j+1,i+1))==1) & ((b(j+1,i)+b(j-1,i)+b(j-1,i+1))==0) ),  a(j,i)= 2^0; break; end, %0
```

```
        if ( ((b(j,i+1) & b(j+1,i))==1)
& ((b(j+1,i+1)+b(j-1,i)+b(j-1,i+1))==0)
), a(j,i)= 2^0; break; end, %0
        if ( ((b(j,i+1) &
b(j+1,i)&b(j+1,i+1))==1) & ((b(j-
1,i)+b(j-1,i+1))==0) ), a(j,i)= 2^0;
break; end, %0
        if ( ((b(j,i+1) & b(j-
1,i+1))==1) & (b(j-1,i)==0) ),
a(j,i)= 2^1; break; end, %1
        if ( ((b(j-1,i) & b(j-
1,i+1))==1) & ((b(j,i-1)+b(j,i+1)+b(j-
1,i-1))==0) ), a(j,i)= 2^2; break;
end, %2
        if ( ((b(j-1,i) & b(j,i+1))==1)
& ((b(j,i-1)+b(j-1,i-1)+b(j-1,i+1))==0)
), a(j,i)= 2^2; break; end, %2
        if ( ((b(j,i+1) & b(j-1,i) &
b(j-1,i+1))==1) & ((b(j,i-1)+b(j-1,i-
1))==0) ), a(j,i)= 2^2; break; end,
%2
        if ( ((b(j-1,i) & b(j-1,i-
1))==1) & (b(j,i-1)==0) ),
a(j,i)= 2^3; break; end, %3
          if ( ((b(j,i-1) & b(j-1,i-
1))==1) & ((b(j+1,i)+b(j+1,i-1)+b(j-
1,i))==0) ), a(j,i)= 2^4; break;
end, %4
          if ( ((b(j,i-1) & b(j-
1,i))==1) & ((b(j+1,i)+b(j+1,i-1)+b(j-
```

```
1,i-1))==0) ),                a(j,i)= 2^4;
break; end, %4
            if ( ((b(j,i-1) & b(j-1,i) &
b(j-1,i-1))==1)   &   ((b(j+1,i)+b(j+1,i-
1))==0) ),  a(j,i)= 2^4; break; end, %4
            if  (  ((b(j,i-1)  &  b(j+1,i-
1))==1)      &       (b(j+1,i)==0)       ),
a(j,i)= 2^5; break; end, %5
            if  (  ((b(j+1,i)  &  b(j+1,i-
1))==1)               &               ((b(j,i-
1)+b(j,i+1)+b(j+1,i+1))==0)               ),
a(j,i)= 2^6; break; end, %6
            if      (       ((b(j,i-1)       &
b(j+1,i))==1)    &    ((b(j,i+1)+b(j+1,i-
1)+b(j+1,i+1))==0) ),       a(j,i)= 2^6;
break; end, %6
            if ( ((b(j+1,i) & b(j,i-1) &
b(j+1,i-1))==1)                           &
((b(j,i+1)+b(j+1,i+1))==0) ),    a(j,i)=
2^6; break; end, %6
           if       (       ((b(j+1,i)       &
b(j+1,i+1))==1)   &   (b(j,i+1)==0)   ),
a(j,i)= 2^7; break; end, %7
           break;
       end
       end
   end
end
%------
%step 3 -   sorting and optimalization
CC = [];
```

```
[y,x] = find(a ~= 0); yx = [y x];
l = 1; o = 0; p = 0;
while (isempty(yx) == 0),
   o = o+1;
   CC(1,o)  =  yx(1,2);      CC(2,o)  =
yx(1,1);                  CC(3,o)          =
log(a(yx(1,1),yx(1,2)))/log(2);
   %start traicing
   j = yx(1,1); i = yx(1,2);
   if (size(yx,1) == 1),
      ContourCollection                     =
AddNewContour(CC,ContourCollection);
      break
   end
   yx              =              [yx(1:l-
1,:);yx(l+1:length(yx),:)];
   if (CC(3,o) == 0),                   i =
i+1; end
   if (CC(3,o) == 1),    j = j-1;     i =
i+1; end
   if  (CC(3,o)  ==  2),      j  =  j-1;
end
   if (CC(3,o) == 3),    j = j-1;     i =
i-1; end
   if (CC(3,o) == 4),                   i =
i-1; end
   if (CC(3,o) == 5),    j = j+1;     i =
i-1; end
   if  (CC(3,o)  ==  6),      j  =  j+1;
end
```

```
   if (CC(3,o) == 7),    j = j+1;     i =
i+1; end
   if (CC(3,o) == 8),    j = j;  i = i;
end      %single pixel
   l = find( (yx(:,1) == j & yx(:,2) ==
i) );
   if      (isempty(l)       ==       1),
%tracing of contour line is finished
      ContourCollection                  =
AddNewContour(CC,ContourCollection);
      l = 1; o = 0; CC = [];
   end
end

function       ContourCollection         =
AddNewContour(CC,ContourCollection),
SP = CC(:,1); CC(:,end+1) = SP;
p      =      size(ContourCollection,1);
ContourCollection{p+1,1} = CC;
```

```
function ContourCollection =
OCE(image);
%
%OCE - Object-oriented Contour
Extraction
%
%image - binary input image
% Author: Dr. Ali Ukasha
ContourCollection = {};

kontur = 1;      tlo = 0;
k = contour;      t = tlo;
%------
%step 1 -   putting zeros to the border
line
[n,m] = size(image);
b = zeros(n+4,m+4);
[aa,bb]=size(b);
b(3:aa-2,3:bb-2) = image(1:n,1:m);
%------
%step 2 -   contour extraction
a = zeros(n,m);
%the table of directional codes
for j = 3:n,
   for i = 3:m,
      if (b(j,i) == 1),
%        while (a(j,i) == 0),
     while (b(j+1,i)==1),
         if ( ((b(j+1,i) & b(j+1,i+1) &
b(j+2,i))==1) &
```

```
((b(j,i)+b(j,i+1)+b(j+2,i+1))==0) ),
a(j+1,i)= 2^0; break; end, %0
        if ( ((b(j+1,i) & b(j+1,i+1)&
b(j+2,i+1))==1) &
((b(j,i)+b(j,i+1)+b(j+2,i))==0) ),
a(j+1,i)= 2^0; break; end, %0
        if ( ((b(j+1,i) &
b(j+1,i+1)&b(j+2,i) & b(j+2,i+1))==1) &
((b(j,i)+b(j,i+1))==0) ),     a(j+1,i)=
2^0; break; end, %0
        if ( ((b(j+1,i) & b(j+1,i+1) &
b(j,i+1))==1) & (b(j,i)==0) ),
a(j+1,i)= 2^1; break; end, %1
%     end
%      while (b(j+1,i+1)==1),
        if ( ((b(j+1,i+1) & b(j+1,i+2)
& b(j,i+1))==1) &
((b(j,i)+b(j+1,i)+b(j,i+2))==0) ),
a(j+1,i+1)= 2^2; break; end, %2
        if ( ((b(j+1,i+1) & b(j,i+1) &
b(j,i+2))==1) &
((b(j,i)+b(j+1,i)+b(j+1,i+2))==0) ),
a(j+1,i+1)= 2^2; break; end, %2
        if ( ((b(j+1,i+1) & b(j+1,i+2)
& b(j,i+1) & b(j,i+2))==1) &
((b(j,i)+b(j+1,i))==0) ),   a(j+1,i+1)=
2^2; break; end, %2
        if ( ((b(j+1,i+1) & b(j,i) &
b(j,i+1))==1) & (b(j+1,i)==0) ),
a(j+1,i+1)= 2^3; break; end, %3
%     end
```

```
%        while (b(j,i+1)==1),
          if ( ((b(j,i+1) & b(j,i) & b(j-
1,i+1))==1) & ((b(j-
1,i)+b(j+1,i)+b(j+1,i+1))==0) ),
a(j,i+1)= 2^4; break; end, %4
            if ( ((b(j,i+1) & b(j,i) &
b(j-1,i))==1) & ((b(j-
1,i+1)+b(j+1,i)+b(j+1,i+1))==0) ),
a(j,i+1)= 2^4; break; end, %4
            if ( ((b(j,i+1) & b(j,i) &
b(j-1,i) & b(j-1,i+1))==1) &
((b(j+1,i)+b(j+1,i+1))==0) ),
a(j,i+1)= 2^4; break; end, %4
            if ( ((b(j,i+1) & b(j,i) &
b(j+1,i))==1) & (b(j+1,i+1)==0) ),
a(j,i+1)= 2^5; break; end, %5

%         end
%          while (b(j,i)==1),
            if ( ((b(j,i) &  b(j,i-1) &
b(j+1,i))==1) &
((b(j,i+1)+b(j+1,i+1)+b(j+1,i-1))==0)
),  a(j,i)= 2^6; break; end, %6
            if ( ((b(j,i) & b(j+1,i-1) &
b(j+1,i))==1) & ((b(j,i-
1)+b(j,i+1)+b(j+1,i+1))==0) ),
a(j,i)= 2^6; break; end, %6
            if ( ((b(j,i) & b(j,i-1) &
b(j+1,i-1) & b(j+1,i))==1) &
((b(j,i+1)+b(j+1,i+1))==0) ),  a(j,i)=
2^6; break; end, %6
```

```
         if ( ((b(j,i) & b(j+1,i) &
b(j+1,i+1))==1) & (b(j,i+1)==0) ),
a(j,i)= 2^7; break; end, %7
         break;
      end
      end
   end
end
%------
%step 3 -   sorting and optimalization
CC = [];
[y,x] = find(a ~= 0); yx = [y x];
l = 1; o = 0; p = 0;
while (isempty(yx) == 0),
   o = o+1;
   CC(1,o) = yx(l,2);   CC(2,o) =
yx(l,1);  CC(3,o) =
log(a(yx(l,1),yx(l,2)))/log(2);
   %start traicing
   j = yx(l,1); i = yx(l,2);
   if (size(yx,1) == 1),
      ContourCollection =
AddNewContour(CC,ContourCollection);
      break
   end
   yx = [yx(1:l-
1,:);yx(l+1:length(yx),:)];
   if (CC(3,o) == 0),              i =
i+1; end
   if (CC(3,o) == 1),   j = j-1;    i =
i+1; end
```

```
   if (CC(3,o) == 2),    j = j-1;
end
   if (CC(3,o) == 3),    j=j-1;        i =
i-1; end
   if (CC(3,o) == 4),                  i =
i-1; end
   if (CC(3,o) == 5),    j = j+1;      i =
i-1; end
   if (CC(3,o) == 6),    j = j+1;
end
   if (CC(3,o) == 7),    j = j+1;      i =
i+1; end
   if (CC(3,o) == 8),    j = j;  i = i;
end      %single pixel
   l = find( (yx(:,1) == j & yx(:,2) ==
i) );
   if (isempty(l) == 1),
%tracing of contour line is finished
      ContourCollection =
AddNewContour(CC,ContourCollection);
      l = 1; o = 0; CC = [];
   end
end

function ContourCollection =
AddNewContour(CC,ContourCollection),
SP = CC(:,1); CC(:,end+1) = SP;
p = size(ContourCollection,1);
ContourCollection{p+1,1} = CC;
```

```
function quit = CESSPCE (A,BW,C,n,m),
% written by: Ali Ukasha
% this program works for all different
images sizes (i.e. square and non
square images)
% CESSPCE is the Contours Extraction
using SSPCE method
% A is the original image
% BW is Black White (binary) image
% C is the Cell array contains all
extracted contours of the image
% n and m the size of the original
image
[I,J]=size(BW);
% start make the image to be squared
if n ~= m,
    mo = (max(size(BW))); n = mo; m =
mo;
end
% finishing the image squared
imagef=zeros(n,m);
for i = 1 : size(C,1),
    CC = C{i,1}; L = size(CC,2);
 image = zeros(n,m); o = zeros(n,m);
   for j = 1:size(CC,2),
      o(CC(1,j),CC(2,j)) = 1;
end
image = image | o;
% za = rot90(image); zz = rot90(za); zx
= rot90(zz);
```

```
% xxz = fliplr(zx);
% imagef = imagef | xxz;
imagef = imagef | image;
end
 za = rot90(imagef); zz = rot90(za); zx
= rot90(zz);
 xxz = fliplr(zx);
 imagef = xxz;

imagef = imagef(1:I,1:J);quit=imagef;
figure; imshow(imagef,[]);
title('Collecting Contours using
SSPCE')
```

```
function quit =
AproxRamer(CC,threshold),
%
%q = AproxRamer(CC,threshold) - Ramer
approximation algorithm
method = 'APP Ramer ALG ';
start_time=cputime;
MSEerror = 0; CompFactor = 0; SNRratio
= 0;
LCC = size(CC,2)-1; L = size(CC,2);
%metoda = 'APP Ramer(P14)';

%first top half
OPEN = []; CLOSED = [];

i = 1; LMP = CC(:,i);
j = find(CC(1,:) == max(CC(1,:))); j =
j(1); RMP = CC(:,j); RMPindex = j;

OPEN(1) = j; CLOSED(1) = j; lC =
length(CLOSED);
OPEN(2) = i; lO = length(OPEN);
while (lO ~= 0),
   i = OPEN(lO); j = CLOSED(lC);
   LiO = CC(:,i); LiC = CC(:,j);
% SL is function that compute the
constants (a, b, c)of the straight line
   abc = SL(LiO,LiC);

   m = CC(:,i+1:j-1); d = 0; dm = 0;
index = 0;
   for k = 1:size(m,2),
```

```
% PD is a function that compute the
perpendicular distance of a point from
the straight line
d = PD(abc,m(:,k));
      if (d > dm), dm = d; index = i+k;
end, end
     if (dm > threshold),
      lO = lO + 1; OPEN(lO) = index;
   else
      lC = lC + 1; CLOSED(lC) =
OPEN(lO); lO = lO - 1; end
end
% WA is the vertices of the contour
WA = CLOSED;

%next bottom half
OPEN = []; CLOSED = []; lCC =
size(CC,2);

i = lCC; LMP = CC(:,i);
j = find(CC(1,:) == max(CC(1,:))); j =
j(1); RMP = CC(:,j);

OPEN(1) = j; CLOSED(1) = j; lC =
length(CLOSED);
OPEN(2) = i; lO = length(OPEN);

while (lO ~= 0),
   i = OPEN(lO); j = CLOSED(lC);
   LiO = CC(:,i); LiC = CC(:,j);
   abc = SL(LiO,LiC);
```

```
   m = CC(:,i+1:j-1); d = 0; dm = 0;
index = 0;
   for k = 1:size(m,2), d =
PD(abc,m(:,k)); if (d > dm), dm = d;
index = i+k; end, end

   if (dm > threshold),
      lO = lO + 1; OPEN(lO) = index;
   else
      lC = lC + 1; CLOSED(lC) =
OPEN(lO); lO = lO - 1;end
end
WA = sort([WA CLOSED]);
%corection of multi-RMP point
NotRMP = find(WA ~= RMPindex);
WA = sort([RMPindex WA(NotRMP)]);   AC
= CC(:,WA);
elapsed_time=cputime-start_time
% AproxMSE is a function that compute
the mean square error between the
original & compressed contours

MSEerror = AproxMSE(CC,WA)/L;
% snr is the signal-to-noise ration &
war is contour variance

snr = 10*log10(war(CC)/MSEerror);
% CompFactor is the contour compression
ratio
CompFactor = 100 * ((L -
length(WA))/L);
```

```
message = strcat(' mse=
',num2str(MSEerror),' snr=
',num2str(snr),' CR=
',num2str(CompFactor));
quit=AC;
figure;plot(AC(1,:),AC(2,:));title(strc
at(method,'
threshold=',num2str(threshold),message)
);
```

```
function abc = SL(A,C),
%
% SL is a function that compute the
constants of the straight line
a = C(2)-A(2);  b = A(1)-C(1);  c =
A(2)*C(1) - C(2)*A(1);  abc = [a b
```

```
function d = PD(abc,B),
% PD is a function that compute the
perpendicular distance of a point from
the straight line
%abc is the straight line constants
%B is the point co-ordinate
d = abs(abc(1)*B(1) + abc(2)*B(2) +
abc(3))/sqrt(abc(1)^2 + abc(2)^2);
```

```
function MSE =
AproxMSE(CC,IndicesOfAproxCC),
%
%q = AproxMSE(CC,IndicesOfAproxCC)
%counts the MSE as sum of squared
distances of points of original contour
points ( CC ) from approximating
polygon ( IndicesOfAproxCC )
%usefull for time domain approximation
methods: Ramer, triangle, tangent,
trapezoid and adaptive-triangle methods

MSE = 0; lA = length(IndicesOfAproxCC);
for i = 1:lA-1,
   iSP = IndicesOfAproxCC(i); iEP =
IndicesOfAproxCC(i+1);

   SP = CC(:,iSP); EP = CC(:,iEP);
   abc = SL(SP,EP);
   for j = iSP:iEP-1,
      MSE = MSE + PD(abc,CC(:,j))^2;
   end
end
MSE = MSE;




function  Y = SNRatio(CC,MSEerror),
```

```
x = CC(1,:); y = CC(2,:);
variance = (var(x,1) + var(y,1))/2;
Y = -10*log10(MSEerror/variance);

function Var_iance = war(CC)
%
%funkcja war(MACIERZ)zwraca wartoœæ
%wariancji macierzy MACIERZ;
%
[n,m]= size(CC);
nm=n*m;
CC = double(CC);
M_CC=(sum(sum(CC)))/nm;  % contour mean
SKL = (CC-M_CC).^2;
Var_iance = (sum(sum(SKL)))/nm;
```

```
function quit =
AproxTrapezoid(CC,th,f),
% - for contours (CC) in Cartesian
representation
% % author: Ali Ukasha ;
% compare the distances of B and C
points from straight lines AC and AD
respectively
method = 'APP Trapezoid';
MSEerror = 0; CompFactor = 0; SNRratio
=0; i=1;
L = size(CC,2);
A = CC(:,i);
B = CC(:,i+f);WA=[1]; C = CC(:,i+2*f);
D = CC(:,i+3*f);

while ((i+3*f) < (L)),
    acS = SL(A,C);
    adS = SL(A,D);
    dC = PD(adS,C);
    dB = PD(acS,B);
if (dB/dC)< th, i=i+3*f;
         if i+3*f > L,break;end
        WA = [WA  i-f];  if i+3*f >
L,break;end

        A=C;B=CC(:,i);C=CC(:,i+f); D
=CC(:,i+2*f);
    else
       i=i+3*f; if i+3*f > L,break;end
```

```
      WA = [WA
i];A=D;B=CC(:,i+f);C=CC(:,i+2*f); D
=CC(:,i);
  end
end
end
WA = [WA  WA(1)];
MSEerror = AproxMSE(CC,WA)/L;
SNRratio = 10*log10(war(CC)/MSEerror);
CompFactor = 100 * ((L -
length(WA))/L);
tekst = strcat(' mse=
',num2str(MSEerror),' snr=
',num2str(SNRratio),' eta=
',num2str(CompFactor));
[MSEerror SNRratio CompFactor NpL];
figure; plot(CC(1,WA),CC(2,WA),'b');
title(strcat(method,' th
=',num2str(th),tekst));
```

```
function quit =
AproxTriangle_Adaptive(CC,th1,f,th2),
% - for contours (CC) in Cartesian
representation
% % author: Ali Ukasha
method = 'APP Triadaptive';
 t1 = tic;
start_time=cputime;ii=1;WA=[];
L = size(CC,2);
A = CC(:,ii);C = CC(:,ii+f);
ii=ii+f;i=1;
  while (ii < (L)),
OPEN = []; CLOSED = [];
LMP = CC(:,ii);
j = find(CC(1,i:ii) ==
max(CC(1,i:ii)));j=ii; j = j(1); RMP =
CC(:,j); RMPindex = j;
OPEN(1) = j; CLOSED(1) = ii; lC =
length(CLOSED);
OPEN(2) = i; lO = length(OPEN);
   i = OPEN(lO); j = CLOSED(lC);
   LiO = CC(:,i); LiC = CC(:,j);
   abc = SL(LiO,LiC);
   m = CC(:,i+1:j-1); d = 0; dm = 0;
index = 0;
   for k = 1:size(m,2), d =
PD(abc,m(:,k));
        if (d > dm), dm = d; index =
i+k; end, end
   if (dm > th1),
```

```
      lO = lO + 1; OPEN(lO) = index;
   else
      lC = lC + 1; CLOSED(lC) =
OPEN(lO);
      lO = lO - 1;   end
%end
SM=sort(OPEN);ln=length(SM);
if ln~=3,WA=[WA   ii];
else
S1=SM(2)-SM(1);S2=SM(3)-SM(2);WA=[WA
SM(2)  SM(3)];
if S1>S2,
EP=CC(:,SM(2));SP=CC(:,SM(1));u1=SM(1);
u2=SM(2);
ff=S1;
else
 EP=CC(:,SM(3));SP=CC(:,SM(2));
 ff=S2;u1=SM(2);u2=SM(3);
end
 LN = CurveLength(CC(:,u1:u2),ff);%LN
is the contour real length
 L0 = sqrt((EP(1)-SP(1))^2 + (EP(2)-
SP(2))^2); );%LO is the direct contour
length


   if ((L0/LN) < th2),
               WA =[WA  u2-round((u2-
u1)/2) u2+round((u2-u1)/2)
];WA=sort(WA);
      SP = CC(:,ii);A=SP;
     if (ii+f)>L,break;end
```

```
      EP = CC(:,ii+f);C=EP;
      else
         WA =[WA  u2-round((u2-
u1)/2)];WA=sort(WA);
         SP = CC(:,ii);A=SP;
         i=ii-f;
         EP = CC(:,ii);C=EP;
   end
end
ii=ii+f;i=ii-f;
end
WA = sort([1 WA  size(CC,2)]);
elapsed_time=cputime-start_time
toc(t1)
MSEerror = AproxMSE(CC,WA)/L;
SNRratio = 10*log10(war(CC)/MSEerror);
CompFactor = 100 * ((L -
length(WA))/L);
text = strcat(' mse=
',num2str(MSEerror),' snr=
',num2str(SNRratio),' CR=
',num2str(CompFactor),' th1=
',num2str(th1),' th2= ',num2str(th2) );
[MSEerror SNRratio CompFactor];
figure; plot(CC(1,WA),CC(2,WA),'b');
title(strcat(method,text));
```

Printed by Books on Demand GmbH, Norderstedt / Germany